Kister Distillation Symposium 2017

Topical Conference at the 2017 AIChE Spring Meeting and 13th Global Congress on Process Safety

San Antonio, Texas, USA
26 – 30 March 2017

ISBN: 978-1-5108-4137-6

Printed from e-media with permission by:

Curran Associates, Inc.
57 Morehouse Lane
Red Hook, NY 12571

proceedings.com

Some format issues inherent in the e-media version may also appear in this print version.

Printed by Curran Associates, Inc. (2017)

For permission requests, please contact AIChE
at the address below.

AIChE
120 Wall Street, FL 23
New York, NY 10005-4020

Phone: (800) 242-4363
Fax: (203) 775-5177

www.aiche.org

Additional copies of this publication are available from:

Curran Associates, Inc.
57 Morehouse Lane
Red Hook, NY 12571 USA
Phone: 845-758-0400
Fax: 845-758-2633
Email: curran@proceedings.com
Web: www.proceedings.com

TABLE OF CONTENTS

482030 AGILE Award Keynote Address

Abstract:

Factors in Selecting AGILE Honoree

1. Overall sustained excellent industry leadership in businesses that employ chemical engineers

2. Recognized for efforts in innovation and creating strategies that ensure success

3. Value of the keynote to the meeting attendees and the members of the Institute Schedule

Event	Time	Location
Continental Breakfast	7:00AM - 8:00AM	Henry Gonzalez Convention Center, Lila Cockrell Theatre
AGILE Keynote Address	8:00AM - 9:15AM	Henry Gonzalez Convention Center, Lila Cockrell Theatre

The 2017 AIChE Spring Meeting and 13th GCPS AGILE Keynote Address will be presented by AGILE award recipient John Y. Televantos, Partner, Arsenal Capital Partners.

John Televantos is a Partner at Arsenal Capital Partners, a middle market private equity firm focused on the specialty industrial, healthcare and specialty financial service sectors. Before joining Arsenal in 2006, Dr. Televantos was President of Aqualon Company, owned by Hercules, Inc.; President and Founder of Helios Chemical Company; and Chief Executive Officer for Foamex International.

Dr. Televantos began his career with Union Carbide in 1977 in silicones research. There he developed two key processes for the manufacture of Methyl Chloride and Siloxane polymers that were commercialized. These processes not only improved efficiency and reduced cost but also dramatically reduced waste. He moved through a variety of management positions there and later at ARCO Chemical, where he became vice president of strategy and development.

At ARCO John co-invented and led a team that developed and commercialized a breakthrough technology in polyol manufacture (a key raw material for polyurethanes) eliminating a whole process step along with waste and emissions, while at the same time opening the door to the creation of new higher-performance polyurethanes. The technology has since been licensed by most of the major global polyol manufacturers with significant waste reduction worldwide.

At Arsenal Capital, Dr. Televantos serves as co-head of the specialty industrial practice. During his tenure, he has been part of several acquisitions and successful exits in the chemicals and materials sectors. In addition to serving on the boards of several Arsenal-owned companies, he is lead director on the board of Balchem Corp.

479518 Tray Jet Flooding – Separating Facts from Fiction

Abstract:

One of the mechanisms limiting the capacity of cross-flow trays is jet flooding. This flooding mechanism is normally associated with high levels of entrainment by either the spray or froth hitting the tray above. Over the years several solutions have been proposed to extend the capacity of cross-flow trays by mitigating this flooding mechanism. Through careful experimentation it was found that some of these solutions do not deliver what is promised. In this presentation data from large scale pilot plants will be used to analyze the underlying fundamentals of jet flooding and possible ways of mitigating it. Feedback from the industry on the efficacy of these solutions will be presented.

477025 Dry Tray Pressure Drop of Sieve Trays Revisted

Abstract:

The sieve tray has been in the market place as a distillation device for many decades. It has been used extensively on distillation tray equipment worldwide as a highly efficient vapor/liquid-contacting device. Many people have examined dry tray pressure drop data for sieve tray decks including both authors. The primary author has published a recent paper correlating the data into an equation. However, he had left the "door open" for additional refinement of this correlation for those applications that might have high tray thickness to hole diameter ratios (e.g. tiny hole applications in aluminum as had been practiced in air separations). The secondary author made two more recent presentations on this very subject employing both simulator data and Computational Fluid Dynamics (CFD) for very thick trays. This current paper will combine all three into a single definitive correlation for the dry tray pressure drop of sieve trays. A correlation for sieve tray pressure drop is presented that incorporates the effects of tray thickness, sieve hole diameter and open area and is compared to data.

Can 2-Pass Moving Valve Trays Experience Multiple Hydraulic Steady States?

By

Matthew Olsson, Fluor, Houston, Texas, USA

and

Henry Z. Kister, Fluor, Aliso Viejo, CA, USA

Presented at the Distillation Topical Conference, AIChE Spring Meeting, San Antonio, Texas, March 26-30, 2017

Summary

Our study reveals that multiple steady state vapor/liquid distributions may be experienced in 2-pass moving valve trays at turndown. Due to the symmetry of 2-pass trays, a perfect split of both vapor and liquid between the passes is always one possible distribution. The others involve having uneven vapor distribution. Some of these steady states effect wide variations in L/V ratios from panel to panel. This maldistribution is likely to lead to major efficiency reductions.

The maldistribution described in this paper is not due to poor distribution of vapor or liquid to the trayed section from an inlet, and it is also not caused by differences in tray geometry between one side and the other, though we have the capability to evaluate these situations.

The cause of the maldistribution is that at turndown, the valves close, and while they are closing, the dry pressure drop remains constant regardless of the vapor rate. This makes it easy for the vapor to swing from one pass to another, forming a number of alternative steady states.

Using our model, we investigated the effect of uniform weeping on the steady state multiplicity. While shifting the steady state some, uniform weeping does not largely impact the steady states. We also looked at the effect of weir loads on the steady state multiplicity. The multiplicity appears to exist both at the lower and higher weir loads. Our findings are that this is a vapor-driven phenomenon with little effect on or influence by liquid unless the L/V ratio is extremely high.

One strategy that appears to alleviate the effects of the multiplicity issue is using valves of different weight. This is a common practice used by valve tray manufacturers. Our study shows that this practice can significantly lessen the severity of maldistribution at turndown, but can also cause a wider range of loadings to experience some maldistribution.

Background

Moving valve trays are known to provide a turndown of 4-5:1 from the maximum operating loads (1,2). However there have been reports (3) and experiences in which turndown of multi-pass trays, including 2-pass trays, fell short of the expected turndown, with indication that maldistribution played a role (3).

In an attempt to gain understanding of the loss in turndown, we applied the Fluor Multi-pass Maldistribution Model (MMM) (4) to some hypothetical tray designs to explore the mechanisms detected. This paper describes our investigation.

Model Tray Design

The tower modeled has 8 ft internal diameter, and contains six 2-pass trays at 24" tray spacing. The top tray, Tray 6, is fed by an internal feed pipe and center false downcomer. The bottom tray, Tray 1, has a sealed center downcomer.

Even trays have side downcomers, and odd trays have center downcomers. The downcomers are straight, with the same width at the top and bottom. The side downcomer width is 11 inches, and the center downcomer width is 10 inches. The clearance between the bottom of each downcomer and the tray deck is 1 inch for both side and center downcomers. 1.5-inch inlet weirs are used to seal the downcomers, and the outlet weirs are 2 inches tall.

The tray deck is 14-gauge. Round moving valves are used, all with 16-gauge thickness and 7/16" leg length. The open slot area is 11% of the bubbling area.

To link with a system that has been studied extensively and for which abundant data and reliable flood and efficiency data and predictions are available, we chose to perform the calculations for the separation of cyclohexane from normal heptane at 24 psia and total reflux. Abundant data for this system in single-pass sieve and valve trays have been published and are available in the literature (5-7).

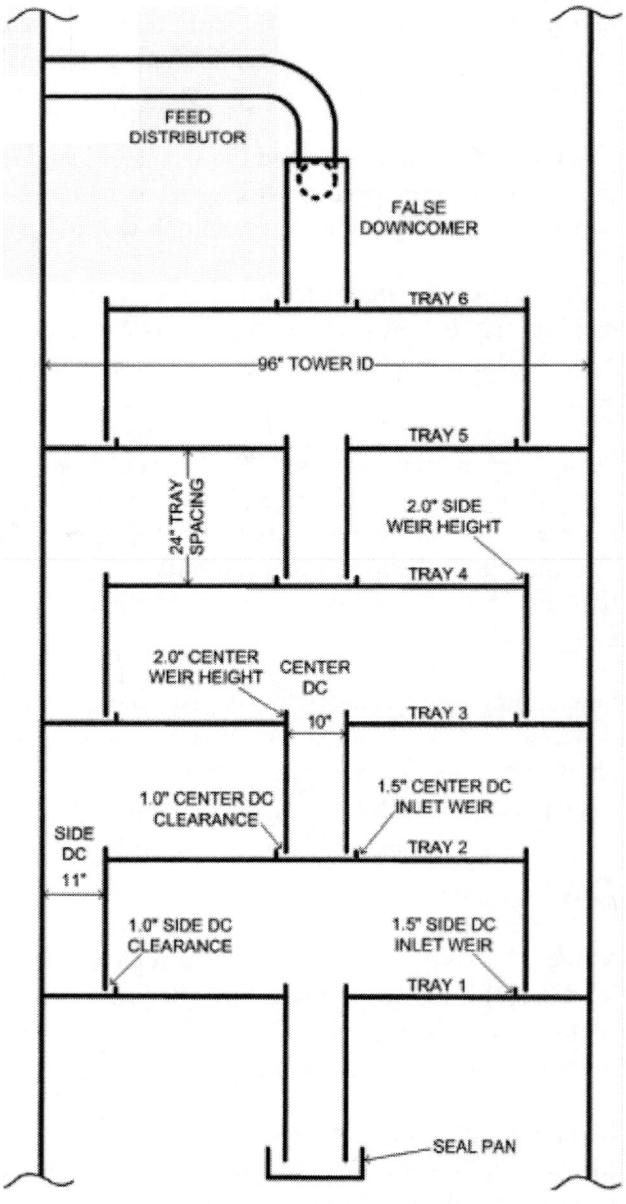

Figure 1. Key Tray Dimensions for Modeled Column.

Modeling Using the Multi-pass Maldistribution Model (MMM)

Our multi-pass maldistribution model is used to evaluate vapor-liquid distributions on multi-pass trays and to find hydraulically balanced distributions. Two equalities are required for a hydraulically balanced distribution for 2-pass trays, and both have to do with shared downcomers. For 2-pass trays, only the center downcomer is shared, and in the modeled column, Trays 1, 3, and 5 have center downcomers as illustrated in Figure 2.

The first equality is pressure drop. On trays for which liquid flows from the center downcomer to the side downcomers, the center downcomer divides the vapor space, and the pressure on either side does not need to be equal. On trays for which liquid flows from the side downcomers to the center downcomer, vapor is continuous above the tray, and the pressure equalizes in the shared space i.e. every two trays. For a balanced solution, the total pressure drop (through both trays) on either side of the shared downcomer must be equal.

The second equality is downcomer backup. As the center downcomer is shared, the calculated downcomer backup for either side must be equal.

The model used for this paper is an extension of our model to calculate maldistribution on 4-pass trays, presented previously (4). The 4-pass model was intended for tray design and rating, dividing the tower in half and hydraulically balancing the asymmetrical passes around the off-center downcomer to find a single solution. The tray design could be refined to match the L/V ratio on either side of the off-center downcomer.

The 4-pass model has only two

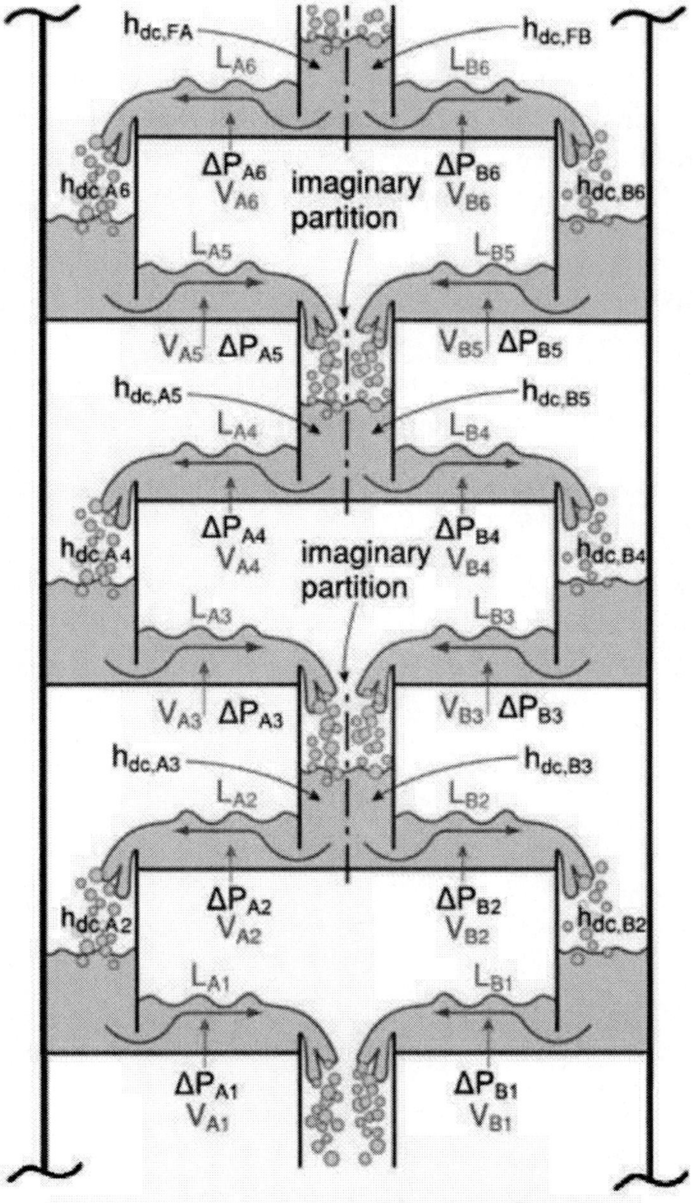

Figure 2. Key terms used in the multi-pass maldistribution model (MMM) equations.

trays which are hydraulically linked, and the liquid rate to and from the center or side downcomer did not change. To obtain thorough insight into valve tray maldistribution, we developed a 2-pass maldistribution model that has 6 trays, hydraulically linked top to bottom, with a liquid feed (false downcomer) to the top tray and a vapor feed to the bottom tray. Total vapor and liquid loadings are constant throughout the tray section for both the 4-pass model and the new 2-pass maldistribution model.

With the 4-pass model, when the vapor loadings were such that one or more tray panels were operating in the region where valves open and close, the maldistribution would be much worse than if all valves were open or all were closed. We recommended avoiding this transition operating region throughout the tray section at design rates.

With 2-pass moving valve trays at turndown, the trays are operating in this transition region. The 2-pass maldistribution tool was designed to search for multiple hydraulically balanced solutions in the transition region. It was also designed to study the effects of feed maldistribution on the tray section, and it has features to calculate weeping and use multiple valve weights on a tray deck.

For the hydraulic calculations, we used the hydraulic equations described in our earlier work (4). For pressure drop calculation, we used the Klein method (8) and the Glitsch method (9), as we have had a lot of success with both. For downcomer backup, we used the classic downcomer backup equation in Perry's Handbook with the Klein method, and the Glitsch downcomer backup equation with the Glitsch method.

The vapor and liquid distributions were closely analyzed at 18%, 22%, 29%, 36% and 40% of the estimated jet flood point.

Steady State Multiplicity

2-pass trays are symmetrical, so a perfectly balanced solution with a 50%/50% split of vapor and the same for liquid is always a hydraulically balanced solution.

However, our MMM analysis found that for 2-pass moving valve trays, at some low loadings there is possibly more than one hydraulically balanced vapor and liquid distribution, a phenomenon we refer to as steady state multiplicity. Such multiplicity shows up when using the Klein pressure drop method (8), but not when using the Glitsch pressure drop method (9). The difference in the behavior of these methods at turndown is due to the combination of the valve tray dry pressure drop characteristic and the aeration factor.

For a moving valve tray, as the gas flow rate increases, the dry pressure drop increases to the point that the valves begin to open, the closed balance point (Figure 3). Then, the dry pressure drop stays constant (for the Klein method) from the gas flow rate when the first valve opens to the rate when all valves are open, the open balance point. Dry pressure drop again increases with the gas flow rate once all valves are open.

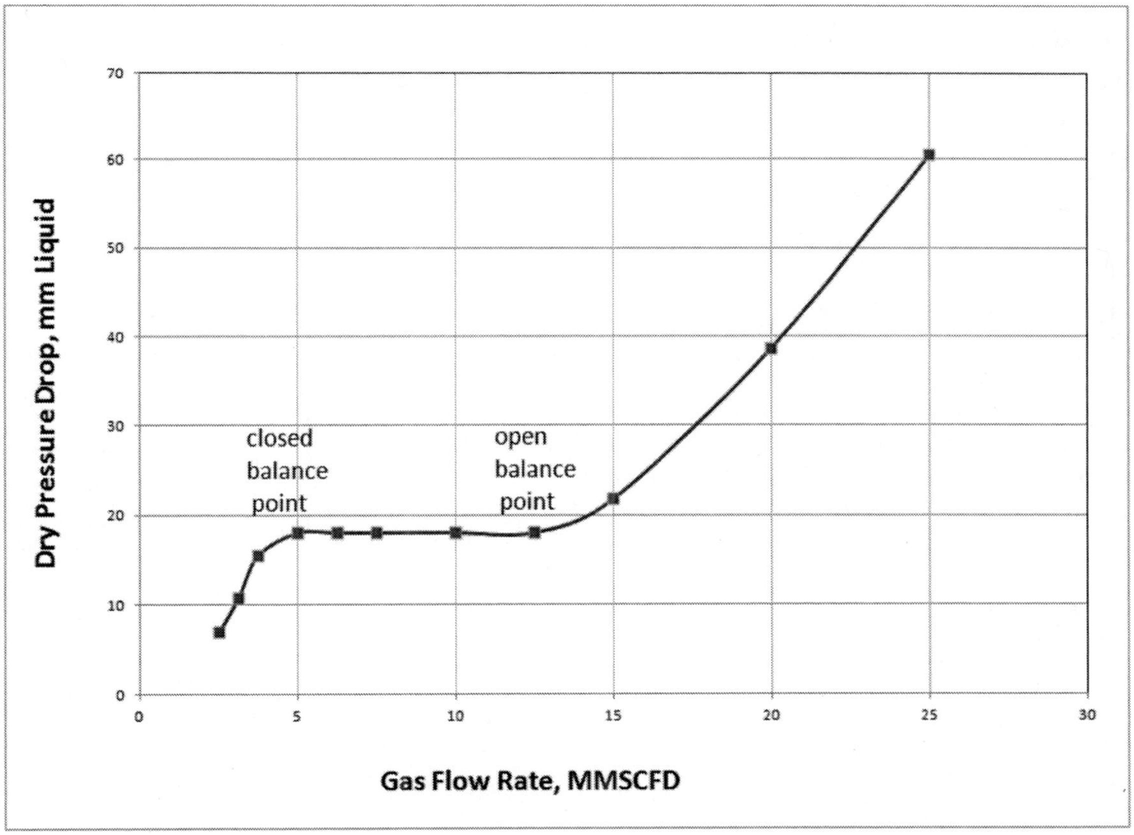

Figure 3. Typical Valve Tray Pressure Drop Characteristic

Multiple Solutions with the Klein Pressure Drop Method

Figure 4 shows the dry pressure drop, pressure drop through liquid, and total pressure drop through Trays 4 and 5 on either side (A and B) of the center downcomer from Tray 5. The loading is 29% of the estimated jet flood point, and the liquid is evenly distributed between passes A and B for both trays. The x-axis is the fraction of vapor to pass A, with the remainder to pass B.

The dashed lines represent the calculated dry pressure drop through both trays for either pass. The pass A dry pressure drop resembles the dry pressure drop characteristic in Figure 3, and pass B is a mirror image. The dotted lines represent the pressure drop through the tray liquid for either pass. The liquid rate remains the same, but per the Klein method, the pressure drop decreases with increasing vapor load as the liquid becomes more aerated.

The solid lines at the top represent the total pressure drop through the trays, the sum of the dry pressure drop and the pressure drop through the liquid. As with the dry pressure drop, the total pressure drop increases with increasing vapor flow rate below the closed balance point and above the open balance point. However, at the rates when the valves open and close, the total pressure drop decreases with increasing vapor flow rate. This is because the dry pressure drop is constant for this region while the pressure drop through the liquid decreases with increasing aeration.

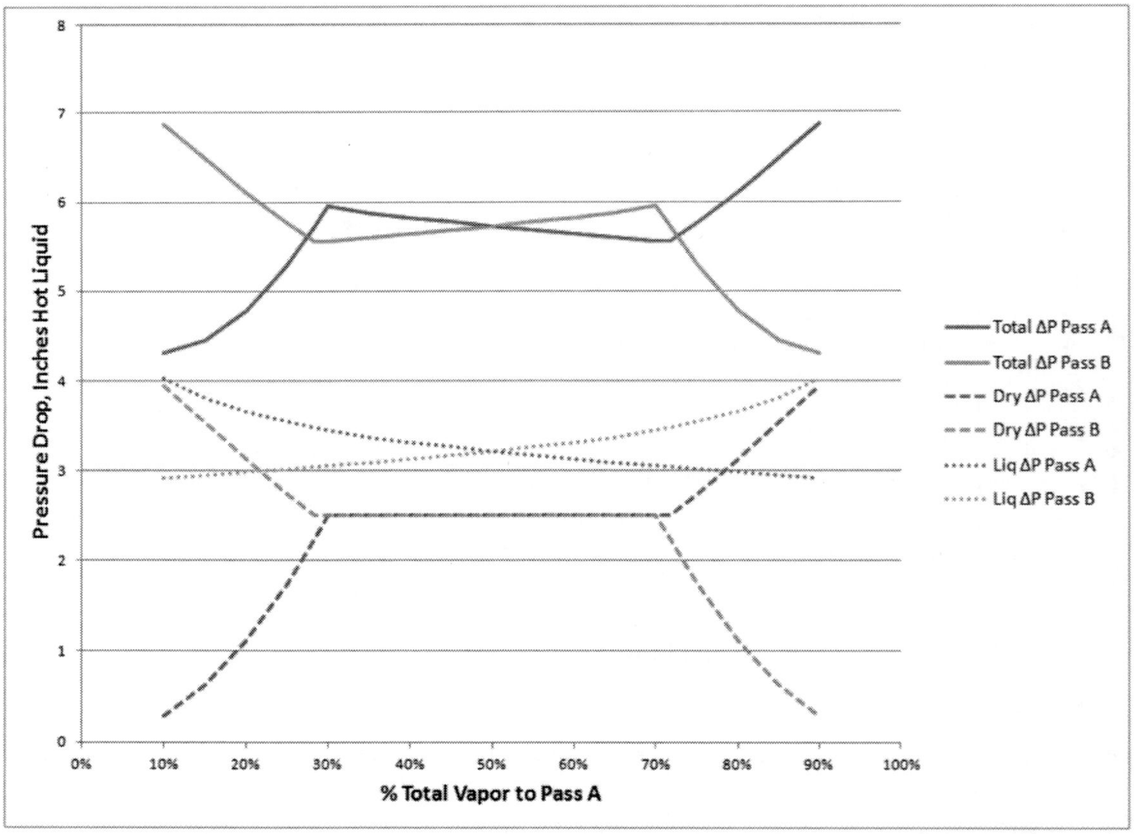

Figure 4. Dry, liquid, and total pressure drop through panels A and B of trays 4 and 5. Klein pressure drop equation, 29% jet flood.

The resulting shape of the total pressure drop curve includes a local maximum at the closed balance point and a local minimum at the open balance point. The total pressure drop curve of pass A intersects that of pass B three times. Each intersection constitutes a potential hydraulically balanced solution, where the pressure is equalized below Tray 4 and again above Tray 5, and the total pressure drop on either side of the center downcomer is equal.

To be a full solution, the downcomer backup in the center downcomer from Tray 5 also needs to be equal for each side. Figure 5 plots the downcomer backup from Tray 5 along with the total pressure drop through Trays 4 and 5 from Figure 4. Their shapes are similar, because the total pressure drop through Tray 5 is a major component of the downcomer backup. The other components (weir height, head over weir, head loss under downcomer, hydraulic gradient) are equal, because the liquid is split evenly between passes A and B for this diagram. The closeness of values of the total pressure drops to the downcomer backups is deceiving, as the plotted pressure drop is the total for two trays (4 and 5), while the downcomer backup is for only one downcomer (from tray 5), so the values of the downcomer backup per tray are of the order of about twice the total pressure drop per tray.

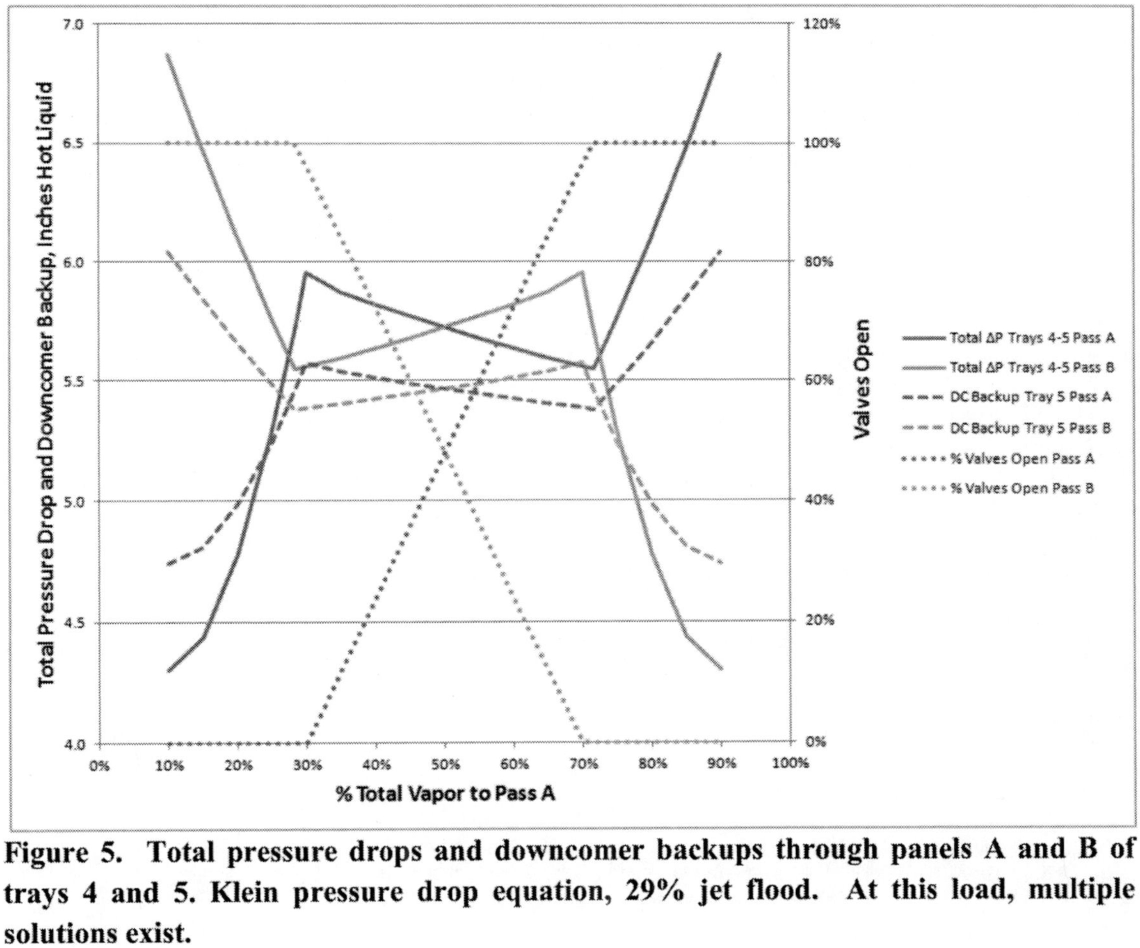

Figure 5. Total pressure drops and downcomer backups through panels A and B of trays 4 and 5. Klein pressure drop equation, 29% jet flood. At this load, multiple solutions exist.

Of course, the solution with both liquid and vapor split evenly between the passes (50% vapor to panel A, 50% liquid to panel A) is a solution on this diagram, as both the pressure drop curves and downcomer backup curves intersect at this point. At about 27% vapor to panel A, and for its mirror image at about 73%, the pressure drop curves and downcomer backup curves each intersect, at nearly the same vapor rate. To line these points up, only a slight liquid maldistribution is needed. For the 27% vapor to panel A, the solution is found with about 50.5% liquid to panel A. For 73% vapor to panel A, the solution is found with about 49.5% liquid to panel A. Each solution other than the even split has this slight liquid maldistribution. Based on our model, the only ways to significantly maldistribute the liquid are either to have a physical difference between the two sides or to have uneven weeping.

Variation of the Multiple Solutions with Percent Jet Flood

Figure 6 is similar to Figure 5, but based on 18% of the estimated jet flood. When the vapor and liquid are evenly distributed, less than 1% of the valves on each panel are open. The only regions present on Figure 6 are the increase in pressure drop with vapor rate when the valves are closed, the local maximum at the closed balance point (at about 50% of the vapor to Panel A or B), and a portion of the region where the valves are opening. Again, multiple solutions are found at the points where all valves are closed on one panel or the other, but, because of the closeness of the evenly split intersection to the closed balance point, the multiple solutions are not much different from each other, so the maldistribution is low. For only slightly less vapor than 18% flood, the evenly split intersection would be the only hydraulically balanced solution because it is in the region where pressure drop increases with vapor rate as all valves are closed.

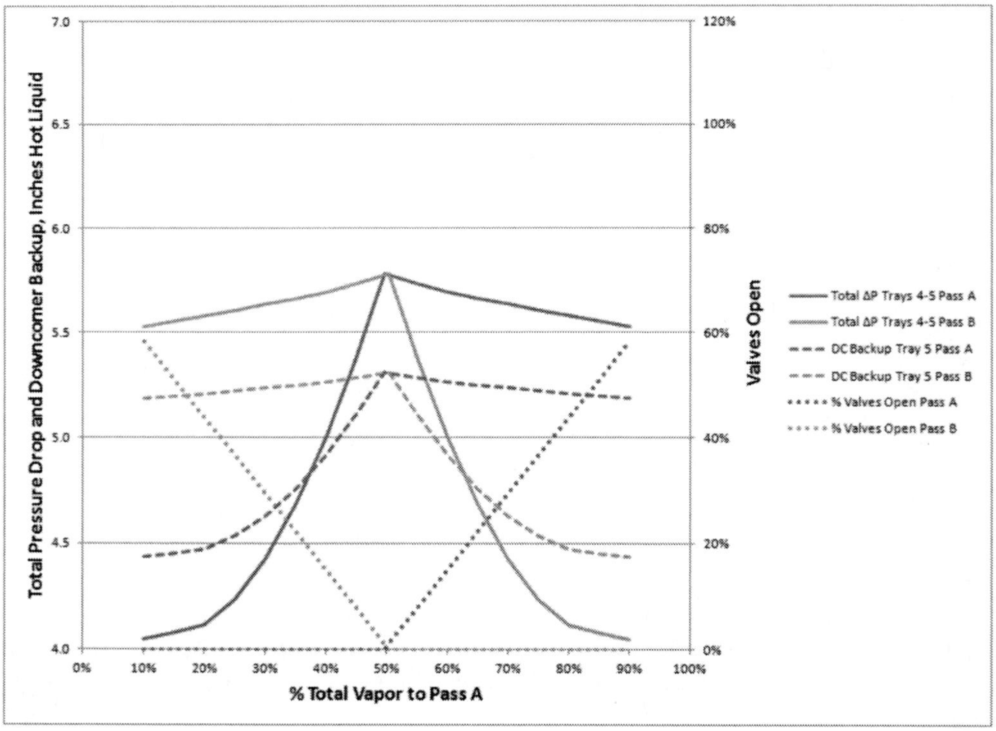

Figure 6. Total pressure drops and downcomer backups through panels A and B of trays 4 and 5. Klein pressure drop equation, 18% jet flood. At this load, multiple solutions still exist, but very close to each other, hardly visible.

For 22% jet flood (Figure 7), over 20% of the valves on each panel open at the evenly split intersection. As with 18% of jet flood, the other solutions are near the points where all valves are closed on one panel or the other. Because of the increased vapor load, these points are farther apart than in Figure 6, so the potential maldistribution at these loadings is greater than for 18% of jet flood.

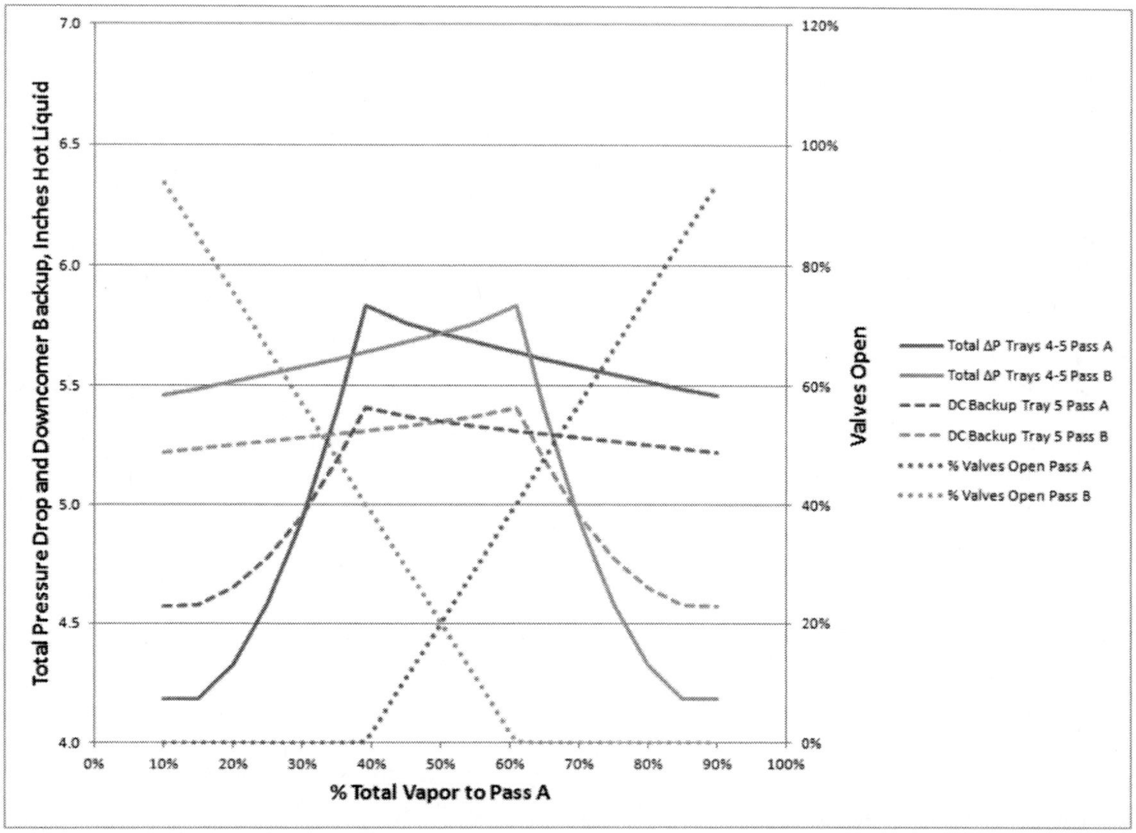

Figure 7. Total pressure drops, and downcomer backups, through panels A and B of trays 4 and 5. Klein pressure drop equation, 22% jet flood. At this load, multiple solutions exist, but closer together than in Figure 5.

For 29% jet flood (Figure 5), nearly 50% of the valves are open at the evenly split intersection, so this is in the center of the pressure drop curve. The other solutions are found at points where all valves are open on one panel and closed on the other. These solutions at 29% of jet flood (27% and 73%) are farther apart than for 22% of jet flood (37% and 63%), so this case has the highest worst-case maldistribution.

For 36% jet flood (Figure 8), about 75% of the valves are open at the evenly split intersection, so this point is closer to the open balance point. The other solutions are at points where all valves are open on one panel and some on the other. Because of the proximity to the open balance point, the multiple solutions are closer together than for 29% jet flood, so it has lower worst-case maldistribution. If the vapor loadings were increased further to the open balance point, the evenly split intersection would be the only hydraulically balanced solution.

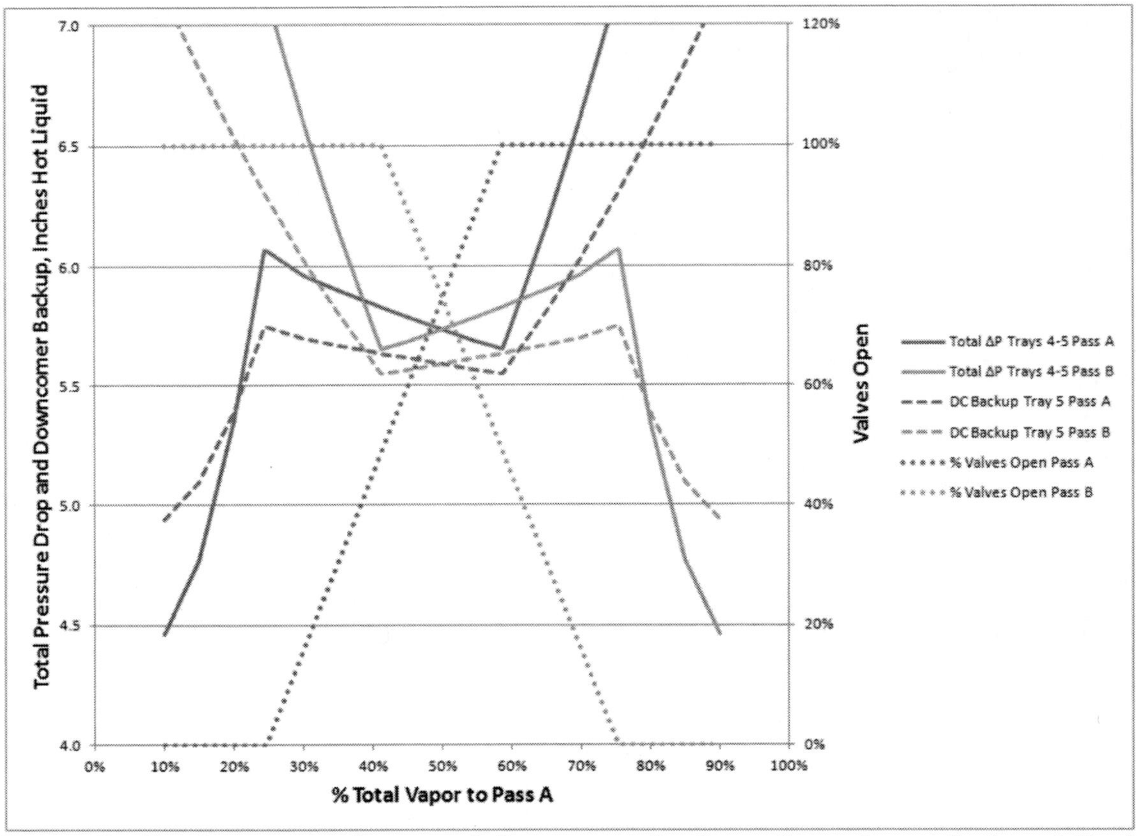

Figure 8. Total pressure drops and downcomer backups through panels A and B of trays 4 and 5. Klein pressure drop equation, 36% jet flood. At this load, multiple solutions exist, and again, closer together than in Figure 5.

The common thread for each of these loads is that at the evenly split intersection, both panels had some valves open but not all, and this is necessary for multiple solutions to exist. Also, the further away from either the open or closed balance point the evenly split intersection falls, the greater the potential maldistribution of the other solutions.

One difference between these loads is how the total pressure drop and downcomer backup for the alternate solutions compare with those for the evenly split solution. For 18% and 22% of jet flood, the maldistributed solutions actually have lower total pressure drop and downcomer backup than the evenly split solution. For 29% jet flood, the pressure drop and downcomer backup for the alternate solutions are slightly lower than for the evenly split solution, as well. For loadings only slightly higher than 29% jet flood, the three solutions would be equal, so switching between balanced solutions may be favored. For 36% jet flood, pressure drop and downcomer backup for the alternate solutions are higher than at the evenly split solution, and this should favor the evenly split solution. Comparing these findings, the implication may be that when greater than 50% of valves are open, even distribution is favored, but when less than 50% of valves are open, maldistribution is favored.

No Multiple Solutions for the Glitsch Pressure Drop Method

The Glitsch pressure drop method does not show multiplicity (Figure 9) because of two differences with the Klein method. The main difference is that for the Glitsch method the aeration factor for the tray liquid is constant, 0.4, so the pressure drop through the liquid does not decrease with increasing vapor load (at constant liquid) as it does with the Klein method. The other difference is that according to the Glitsch method, the dry pressure drop slightly increases with vapor load in the region where valves open and close, as opposed to the Klein method, where the dry pressure drop stays flat in that region (as in Figure 3).

Combining these two factors, the total pressure drop as calculated by the Glitsch method always increases with increasing vapor load. This results in only one intersection between the pressure drop curves of pass A and pass B, which is always at the evenly split distribution.

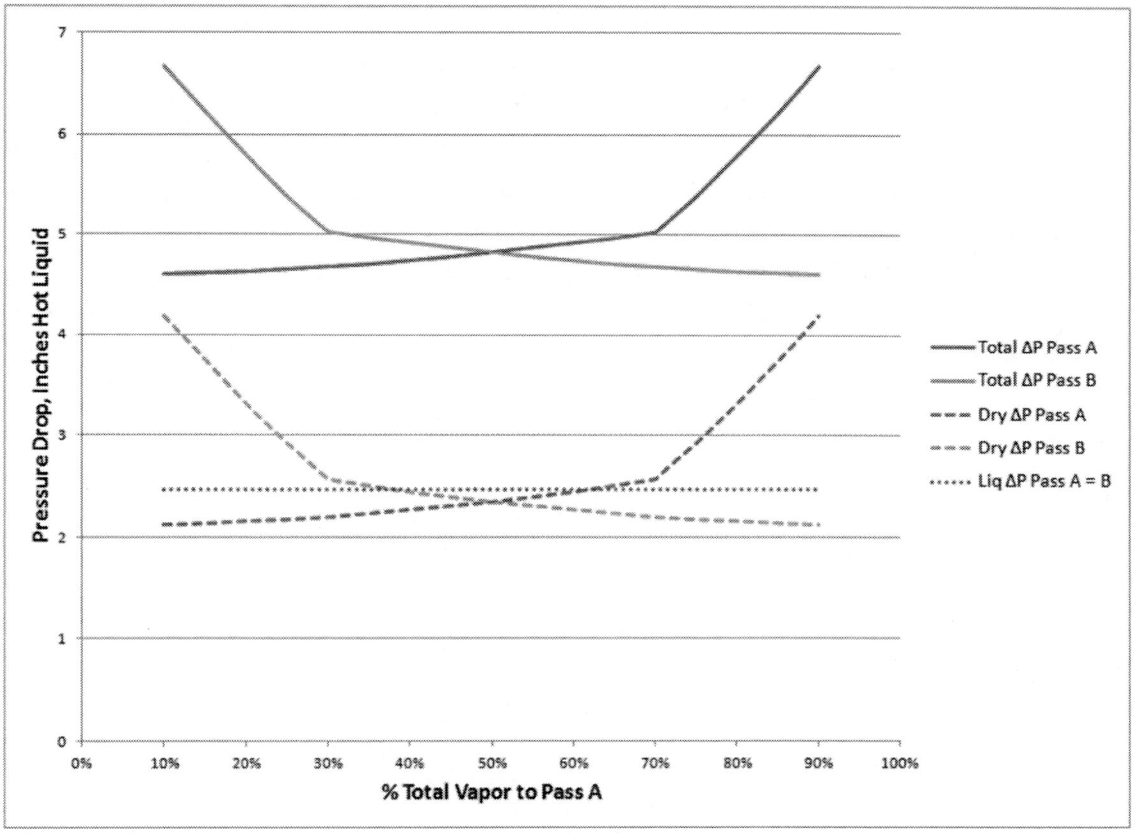

Figure 9. Dry, liquid, and total pressure drop through panels A and B of trays 4 and 5. Glitsch pressure drop equation, 29% jet flood.

Which Correlation is Better Here: Klein or Glitsch?

This brings the question: Which correlation do we believe? The key to the answer is to appreciate that both correlations focus on the region in most important to designers, i.e., when the valves are open and pressure drop is high. This is the region where dry pressure drop dominates. In this region both correlations predict much the same and their trends are robust and reliable.

Neither correlation was intended to address distribution in multi-pass trays at turndown. At turndown, measured pressure drops are low and subject to greater errors, so it is likely that the pressure drop trends are inaccurate. Further, due to the flatness of the pressure drop curve in this region, instability is likely, and gas can flip-flop from one panel to another, similar to the instability described by Brierley et al. (10).

An argument supporting the suitability of the Klein correlation for this analysis is that there is considerable evidence to suggest that due to increased aeration the liquid pressure drop on a tray diminishes as the vapor load is raised. This is featured in the Bolles and Fair correlation for sieve trays (11), the Colwell clear liquid height and froth density correlation (12), and even by a control issue reported by Buckley et al. (13). The Klein method's approximation of a liquid head that diminishes with an increasing vapor flow rate appears much more realistic than the Glitsch method's approximation that the aeration factor for the tray liquid is constant, 0.4, regardless of the vapor load. We therefore proceed on the basis of the Klein pressure drop correlation.

Distribution Criteria

Tray Stability

For low vapor loads, a criterion termed *tray stability factor*, η, was proposed by Summers et al. (3). It is a variation of the Froude number, comparing the upward inertial force of vapor to the downward gravitational force of liquid.

$$\eta = (\Delta P_{DRY}/H_{CL})^{0.5}$$

ΔP_{DRY} = calculated dry tray pressure drop, inches of liquid

H_{CL} = hydrostatic head, inches of liquid = total pressure drop – ΔP_{DRY}

Summers found that stability factors exceeding 0.6 limit weeping to below 30% and render trays stable. Note that Summers' method (3) derived the hydrostatic head H_{CL} values from total pressure drop measurements by subtracting from them calculated dry pressure drops as defined above. In the absence of measured total pressure drop data, Summers (14) recommends using the Colwell clear liquid height correlation (12) for calculating stability factors, and reports excellent experience with it. The values for the stability factors for our model runs are plotted on Figure 10 below. Figure 10 shows that the stability factors for the modeled trays were above 1.05 down to 20% of jet flood, suggesting no instability, less than 30% weeping, and minimal adverse effects due to weeping.

Summers et al. (3) also proposed an additional crude stability criterion for 2- and multi-pass trays to determine whether the total vapor loading is sufficient to make it likely that vapor will flow through each tray pass. The criterion is whether:

$$\Delta P_{ACTIVE} > \Delta P_{INACTIVE}$$

For this test, liquid would be distributed evenly among the tray passes. The inactive pass would have no vapor loading, and the full vapor loading would be divided among the remaining passes. In the case of 2-pass trays, all vapor would travel through one pass, with the liquid split evenly. The pressure drop through the inactive side is the sum of the weir height and the head over the weir, using the Francis Weir Formula (2). The pressure drop through the active side is the sum of the dry pressure drop and hydrostatic head calculated for the aerated liquid from Colwell (12). An equivalent expression for the multi-pass test is:

$$\Delta P_{ACTIVE}/\Delta P_{INACTIVE} > 1$$

We applied this test to the loadings used in our model and added the results to Figure 10. As long as the $\Delta P_{ACTIVE}/\Delta P_{INACTIVE}$ ratio is >1, this Summers et al. criterion expects no additional issues due to maldistribution between the passes. Figure 10 shows that this ratio exceeds unity for loadings above 25% of jet flood and is below unity at lower loadings.

Overall, of the two Summers et al. stability criteria, one (multi-pass instability) is consistent with our findings of 2-pass valve tray maldistribution at turndown, but only below 25% jet flood. The other (tray instability) does not anticipate this maldistribution. Neither anticipates the modeled maldistribution in the range of 25% to 40% of flood as is discussed below.

Bolles Distribution Ratios

Maldistribution varies the liquid to vapor (L/V) ratio from pass to pass. Pinching, and therefore poor separation, occurs in the low L/V regions, while the high L/V ratios in other regions only slightly offset the separation loss. Overall, observed efficiencies diminish, sometimes grossly. Bolles (15) defined the distribution ratio Φ as the ratio of the maximum pass L/V to the minimum pass L/V, and recommended keeping this distribution ratio below 1.2 for good tray efficiency. Summers (16) tightened Bolles' criterion to 1.1.

Distribution ratios can readily be calculated from our MMM model, since our model calculates the L and V for each pass for a given hydraulically balanced solution. Figure 10 shows the distribution ratios for the "worst-case maldistribution" from the MMM for each percent of jet flood based on the Klein pressure drop method. The term "the worst-case maldistribution" is used because we found up to three hydraulically balanced solutions loadings up to the vapor rate where all valves are open. One of the three is an evenly split distribution (distribution ratio of 1). The other two (mirror images in Figures 5 to 8) both have the same high "worst-case maldistribution" distribution ratios.

Figure 10 shows that for the tests between 20% of flood and 40% of flood, the distribution ratios were well above the maximum of 1.2 specified by Bolles, reaching a huge ratio of about 2.6 at 30% of flood. We have seen significant losses in tray efficiency in multi-pass trays when the distribution ratios exceeded 1.3, so 2.6 is huge.

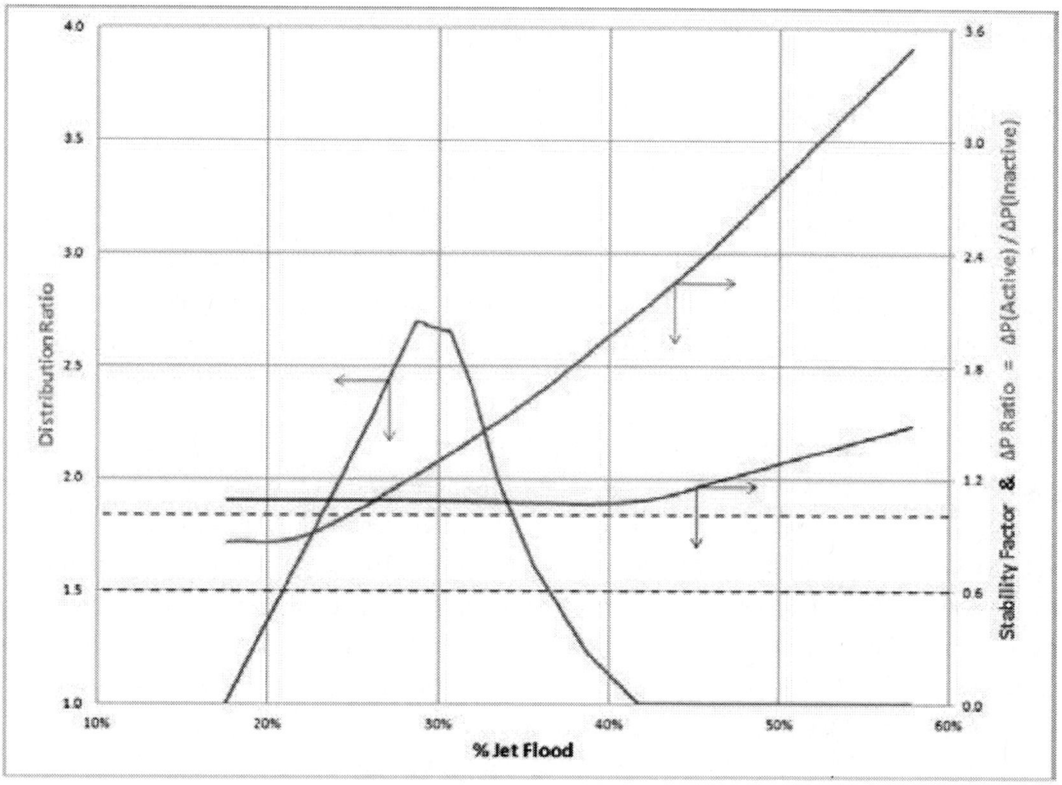

Figure 10. Applying the Summers et al. hydraulic metrics to explore instability: Stability Factor (red curve) and ΔP Ratio (green curve), compared with Bolles Distribution Ratio for Worst-Case Maldistribution calculated from our MMM model (blue curve).

What affects the Steady State Multiplicity?

Multiple Valve Weights

Tray manufacturers to often use two different valve weights on each tray panel, typically with 50% light gauge valves and 50% heavy gauge valves (17), to improve efficiency at turndown. The different weights are evenly spread across the tray (typically one row of lights, one row of heavies, rows perpendicular to the liquid flow) with the light valves forming the front row facing the incoming liquid. We explored the efficacy of this strategy to alleviate steady state multiplicity.

The dry pressure drop characteristic of a two-weight valve tray is similar to that of a single weight valve tray, except that there are two flat regions instead of the one in Figure 3, one for each valve weight. On Figure 11, for 29% flood, as the gas flow rate increases, the dry pressure drop increases to the point that the light valves (16-gauge) begin to open, the *light closed balance point*. Then, the dry pressure drop stays constant to the rate when all light valves are open, the *light open balance point*. Dry pressure drop again increases with the gas flow rate once all light valves are open, up to the point that the heavy valves (14-gauge) begin to open, the *heavy closed balance point*. Then, there is a second region in which the dry pressure drop is constant, up to the gas rate that all valves are open, the *heavy open balance point*. Above this point, dry pressure drop increases with increasing gas flow rate.

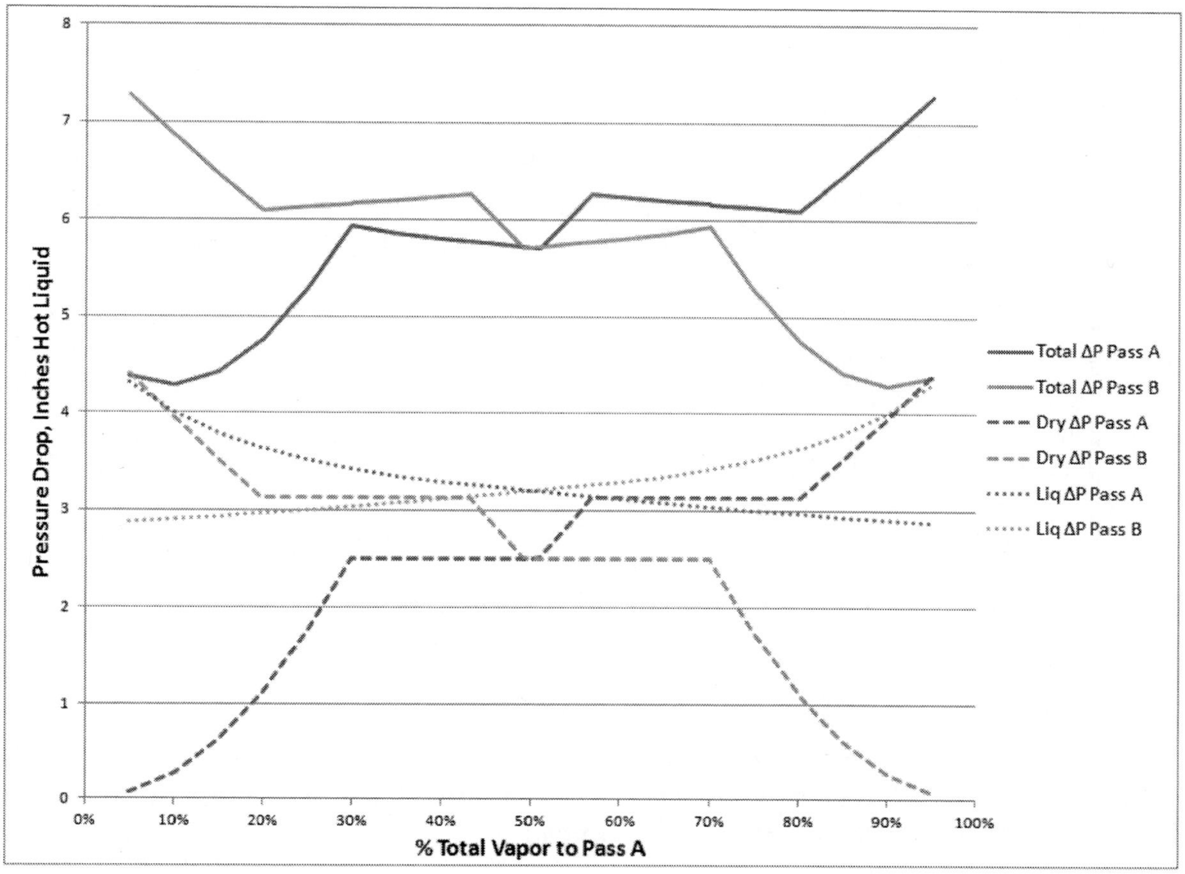

Figure 11. Dry, liquid, and total pressure drop through panels A and B of trays similar to trays 4 and 5, but with two weights of valves, 14-gauge and 16-gauge. Klein pressure drop equation, 29% jet flood.

Figure 11 shows that for 29% jet flood, a two-weight design would nearly eliminate the potential maldistribution seen in Figure 4. However, Figure 13 shows that at 40% of jet flood loadings, which had very little potential maldistribution with a single valve weight (Figure 12a), a two-weight valve design could experience significantly more maldistribution (Figure 12b).

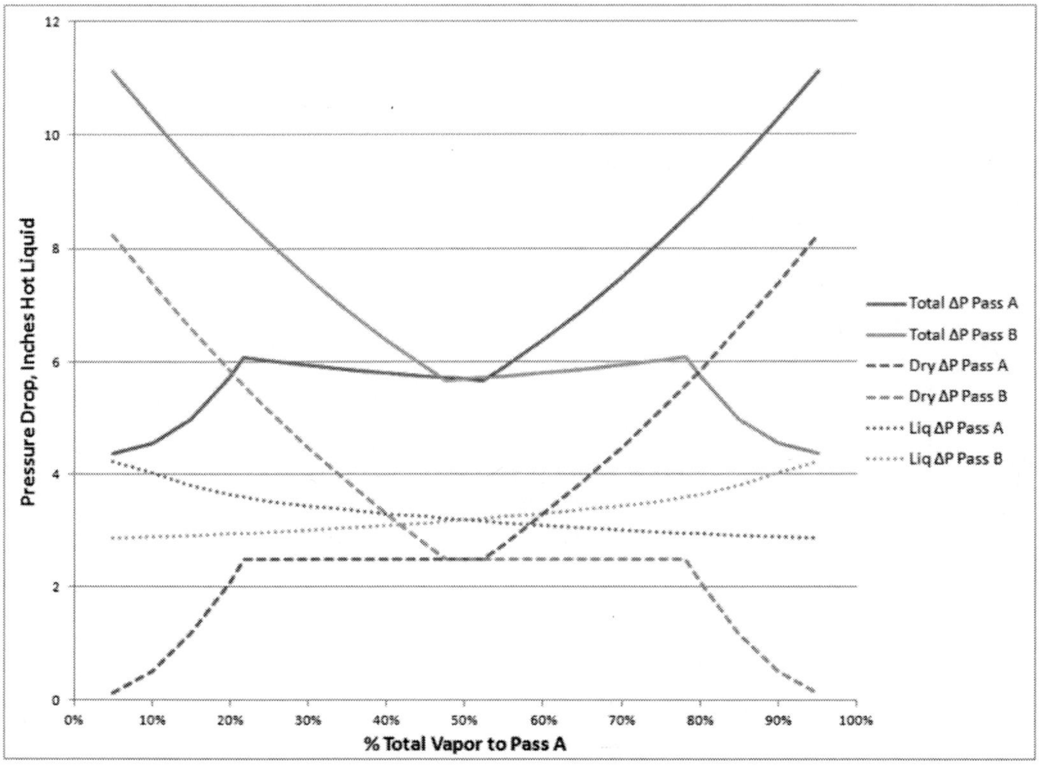

Figure 12a. Dry, liquid, and total pressure drop through panels A and B of trays 4 and 5. Klein pressure drop equation, 40% jet flood.

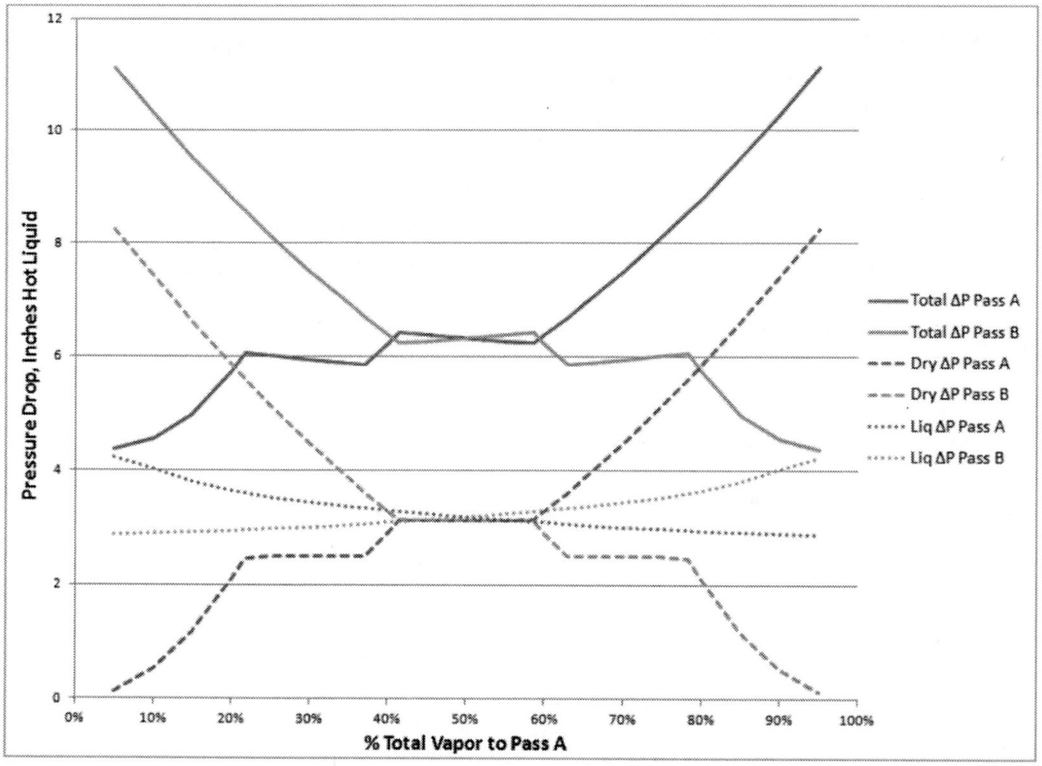

Figure 12b. Dry, liquid, and total pressure drop through panels A and B of trays similar to trays 4 and 5, but with two weights of valves, 14-gauge and 16-gauge. Klein pressure drop equation, 40% jet flood.

Figure 13 compares the distribution ratios for the worst-case maldistribution for the single valve weight to those of similar trays containing 50% light (16-gauge) and 50% heavy (14-gauge) valves. For the two valve weights the distribution ratios are much smaller, indicating a tremendous improvement. The turndown also improved, with the major maldistribution peak moving from about 30% to 23% of jet flood. On the debit side, the opening of the heavy valves generated a new maldistribution peak centered around 40% of jet flood. However, at this peak the maximum distribution ratio was 1.4, much less than the maximum in the single weight valve case, and in half of that peak the distribution ratios were within the Bolles criterion of 1.2 maximum. So the efficiency loss due to that peak should be much less than with the single-weight valves.

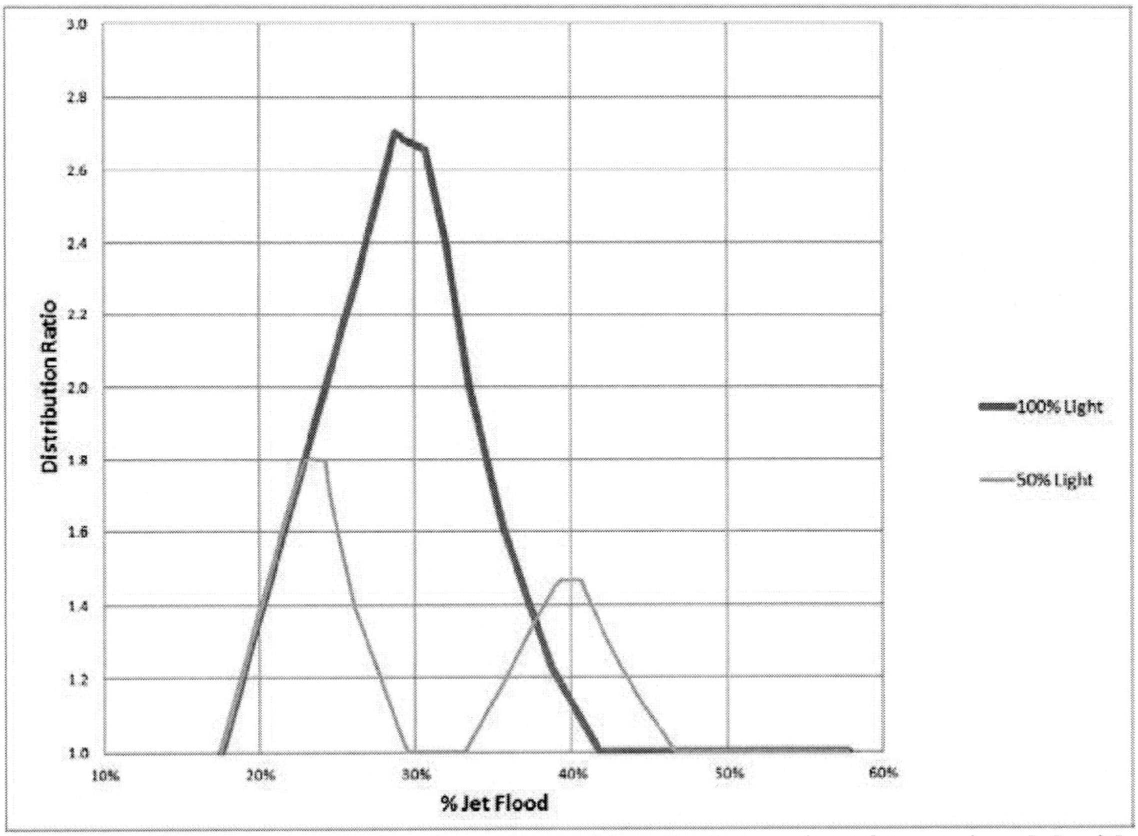

Figure 13. Comparing calculated worst-case distribution ratios of trays 4 and 5 with uniform valve weights (blue) to the calculated worst-case distribution ratios with two weights of valves, 14-gauge and 16-gauge (green).

Heavy Valve Gauge

To explore the effect of valve thickness, we ran the model again, pairing the 16-gauge valves with 12-gauge valves (Figure 14).

Comparing Figure 13 to 14, the distribution ratios are quite similar, and both are much lower than for the single weight valves as described earlier. Compared to the 14-gauge/16-gauge valves, the 12-gauge valves open at higher vapor rates than the 14-gauge valves. The maldistribution peaks associated with the heavy valves closing are shifted away from those of the light gauge valves. This adversely affects distribution closer to design rates, but it increases the range of good distribution between the peaks, where light valves are open and heavy valves are closed. There is no difference between the figures at vapor rates where only light valves are open.

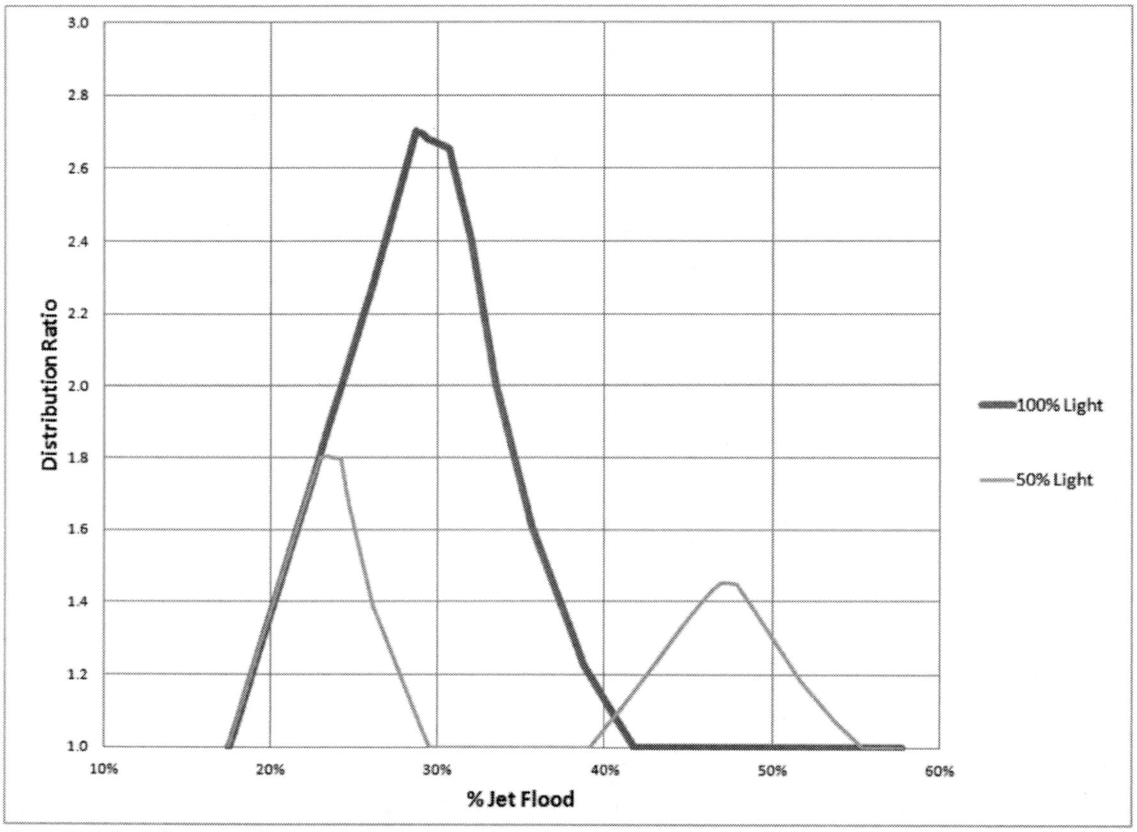

Figure 14. Comparing calculated worst-case distribution ratios of trays 4 and 5 with uniform valve weights (blue) to the calculated worst-case distribution ratios with two weights of valves, 12-gauge and 16-gauge (green).

Multiple Valve Weight Strategy

Figures 13 to 14, show that employing a multiple valve weight strategy tremendously reduces the distribution ratios for the worst-case maldistribution, and therefore has a beneficial effect at turndown. However, some loss of efficiency may be encountered at higher loads due to the opening of the heavier valves which is encountered at higher percent of jet flood value. Since it is also common that loadings vary significantly throughout a trayed section, while the tray design does not, at any point in time different trays in the tower may operate at different points on the same diagram. Whatever the case, the valve weight strategy should be reviewed for consistency with the predicted tower operation.

Vapor Tunnels

One strategy that is sometimes employed for multi-pass trays to allow for vapor to redistribute based on the pressure drop through each tray is to use vapor tunnels (4, 16). Vapor tunnels are passages through a shared downcomer which allow vapor to pass based on pressure differential between the tray passes on either side of the downcomer. These may also equalize the clear liquid height on the tray being fed from the shared downcomer, depending on their location vertically in the downcomer.

For this 2-pass trays case, vapor tunnels through the center downcomer would equalize the pressure on either side of the center downcomer. This would decouple the vapor distribution of the tray with the center downcomer from the distribution to the tray below. This would not address the cause of maldistribution at turndown, which is that the vapor rate is insufficient to open all the valves on the tray decks, resulting in no dry pressure drop difference between the evenly split solution and the maldistributed one. The difference would be that with vapor tunnels, every tray would have its own vapor distribution, rather than every two trays, and by this reasoning vapor tunnels should not significantly change the performance of a trayed section.

Weir Loads

Since our analysis was based on the cyclohexane-normal heptane system at 24 psia and total reflux, the weir loads were low. For any of the loadings between the closed balance point and the open balance point, the weir loading was less than about 2 gpm/inch for the side downcomers and less than about 1.5 gpm/inch for the center downcomers.

To determine if the multiplicity would persist at higher liquid loadings, we evaluated the turndown behavior of an alternate case. The alternate case has the same internal diameter, downcomer dimensions, and valve type as those in our model. The difference is that the weir loading was raised to 8 gpm/inch at 80% of flood. The L/V ratio for this case was 2.4, much higher than the L/V ratio of 1.0 for the total reflux case. An L/V ratio of 2.4 is of the order used in a stripping section of a fractionator, where 2-pass trays are frequently used.

The increase in liquid rate and L/V ratio did not increase the influence of the liquid on maldistribution. The maldistribution chart for the alternate case, Figure 15, looks the same as the one for the lower weir loads in Figure 13. The peaks appear to have shifted to the right. This is because for a given vapor rate, the increased liquid rate increased the percent jet flood. The vapor rates at which valves opened and closed did not change from Figure 13.

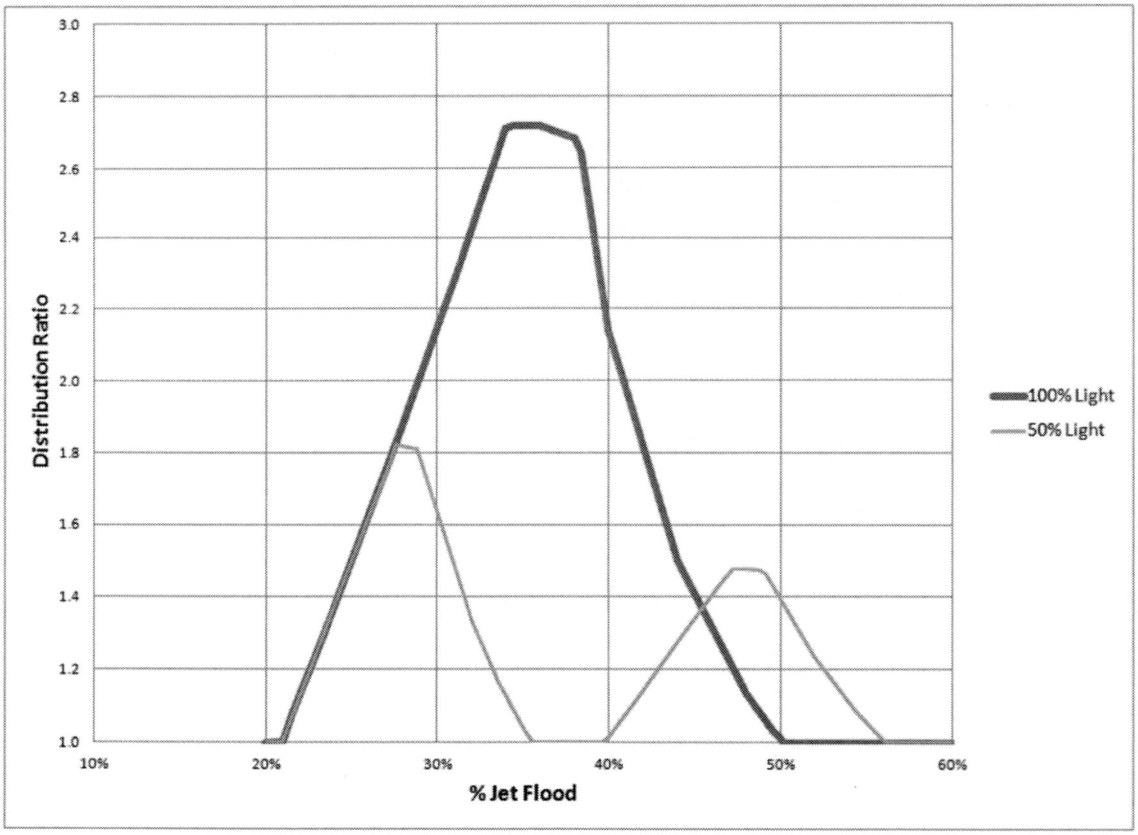

Figure 15. Comparing calculated worst-case distribution ratios between trays 4 and 5 for alternate loadings with higher weir loading than the total reflux, single weight 16-gauge valves (blue) vs. two weights of valves, 14-gauge and 16-gauge (green).

Unique Multiplicity at Extremely High L/V Ratios

For the alternate case in Figure 15, the L/V ratio was increased such that the weir loading would be 8 gpm/in at 80% of flood, and there was not a significant change in behavior from the total reflux case. Indeed, this is true for even higher weir loadings as well. For these cases, using multiple valve weights (16-gauge and 14-gauge) would be a good solution to limit maldistribution at turndown.

However, in some cases, the L/V ratio can be increased further, well above the normal range, such that the limiting factors in tray design are downcomer choke flood and downcomer backup flood rather than jet flood. We evaluated a second alternate case with a much higher L/V ratio to determine whether such liquid-dominated loadings would experience any unique behavior.

For the second alternate case, we used an L/V ratio of 8.6, much higher than the first alternate case (2.4) or the total reflux case (1.0). Unlike for the previous alternate case, the tray geometry needed altering for this high L/V case. The downcomers employed for the lower L/V ratios were far too small for the 8.6 ratio. To achieve a comparable design, the diameter of the tower was increased from 96" to 108", with all the additional area added to the downcomers, maintaining straight downcomers. The active and the valve open slot areas were unchanged. The side downcomer width was increased from 11" to 17.3", and the side downcomer clearance was increased from 1" to 4". The center downcomer width was increased from 10" to 17.75", and the center downcomer clearance was increased from 1" to 3".

With this design and 8.6 L/V ratio, at the vapor rate for which about half the valves are open (29% jet flood for total reflux loadings and modeled column trays), the weir loading is 9.3 gpm/in, and the trays are at 39.3% jet flood. As the loadings are increased to 80% jet flood, the weir loading is 19 gpm/in, the downcomer inlet velocity is just below 0.51 ft/s, and the center downcomer backup is 12.9" clear liquid. These values push against the maximum good practice limits (2), but do not violate them, giving at least a reasonable design for much of the operating range.

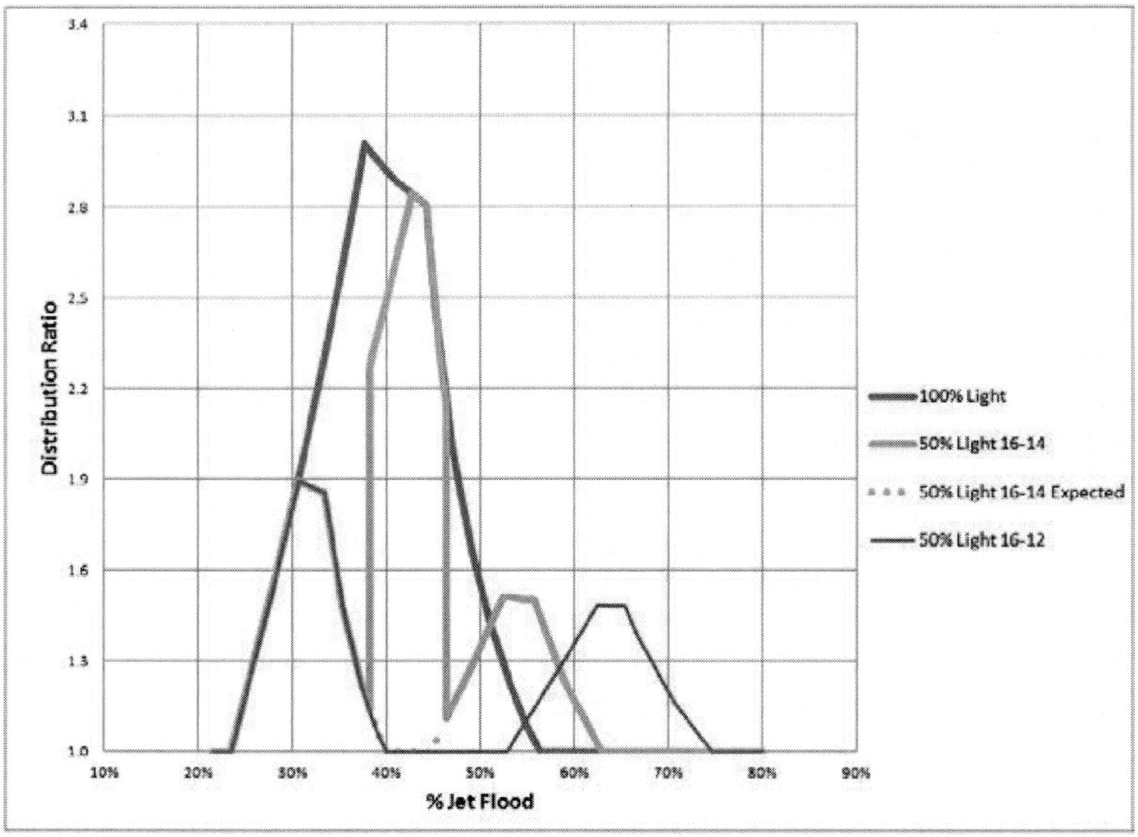

Figure 16. Comparing calculated worst-case distribution ratios between trays 4 and 5 of the larger diameter tower with high L/V loadings, single weight 16-gauge valves (blue) vs. two weights of valves, 14-gauge and 16-gauge (green) or 12-gauge and 16-gauge (red).

The distribution ratios for the 16-gauge single valve weight design and for two multiple valve weight designs (16-gauge/14-gauge & 16-gauge/12-gauge) are shown in Figure 16. For the single valve weight design, as well as for the 16-gauge/12-gauge design, the distribution ratio curves are similar to those for the lower L/V ratios (compare to Figure 14). In contrast, the design using 16-gauge and 14-gauge valves appears to no longer reduce peak maldistribution as effectively as it does for the lower L/V cases.

The portion of the distribution ratio curve for the 16-gauge/14-gauge design which deviates from the other two-weight curves is between 38% flood and 46% flood. The conventional solutions are still possible and shown as the dotted line, but for this range there are more maldistributed solutions. For the portion that the curve shares with the peak of the single-weight curve, all light and heavy valves are open on panel A, and all valves are closed on panel B.

For each of the designs discussed in this paper, the liquid is not maldistributed more than 0.5-1%, and that is also the case for the 8.6 L/V ratio. The difference in pressure drop through liquid for panel A and panel B is then due to the difference in the aeration factor, based on the amount of vapor maldistribution. For lower L/V ratios, the difference in pressure drop through liquid at turndown is very small, as on Figure 13, and is only significant enough to lead to an alternate solution for a single-weight design. With an 8.6 L/V ratio, the difference in pressure drop through liquid becomes much more pronounced. With the relatively small difference between the dry pressure drop at which the 16-gauge and the 14-gauge valves open, the more pronounced difference in pressure drop through liquid can exceed the dry pressure drop difference between the opening of the 16-gauge and 14-gauge valve weights. This allows for the behavior where heavy valves can be open on panel A, while light valves are closed on panel B. This is unique behavior for very high L/V ratios.

In order to eliminate this behavior, it is possible to increase the difference in dry pressure drop between the valve weights such that it is greater than the difference in pressure drop through liquid (due to aeration). This can be done by using 16-gauge and 12-gauge valves, shown in Figure 16.

Weeping

We estimated for each pass the amount of liquid that weeps through the valves on the tray deck and bypasses the downcomer. To do this, we used the Colwell and O'Bara modification (18) of the Lockett and Banik correlation for sieve trays (19), found in *Distillation Design* (20). The correlation uses the Colwell clear liquid height (12), which is a function of the weir load. The weeping liquid was removed from the weir load, and an iterative solution needed to be performed for each tray pass.

As this is a sieve tray correlation, it uses the full open area. This produced results with over 80% of tray liquid weeping for 18% flood and over 40% of tray liquid weeping for 22% flood. These results are unrealistic for moving valve trays because the valves can close, lowering the open area and increasing the vapor velocity, two factors which reduce weeping.

To correct for this, the open area was adjusted based on the number of valves open on the tray. For the Klein correlation, the closed balance point and open balance point are calculated in terms of the vapor velocity through the fully open hole area, with the closed balance point at a lower velocity and the open balance point at a higher velocity. The number of open valves is found by interpolating between these two points. The open area for weep is then adjusted based on the number of valves open. Closed valves are not completely closed, however. As an approximation, the velocity ratio between the open balance point and the closed balance point was used as the area ratio between an open valve and a closed valve.

With the area correction in place, the weeping was limited to less than 5% of tray liquid for all cases, making it mostly irrelevant to the distribution behavior discussed in this paper. It is possible that this is an overcorrection and that more weeping is possible at turndown. This would especially be true in the region where valves are opening and closing with no effect on the dry pressure drop, as some valves could be open for vapor while others are open but mostly weeping. This path was not explored as our model does not have a method to find solutions with this behavior, other than arbitrarily.

Uneven Weeping

We also explored the effects of uneven weeping on vapor and liquid distribution. We specified that for one side of the tower, 20% of the tray liquid would weep, while for the other side, 0% would weep. As the same amount of liquid weeps on each tray, the weeping liquid hydraulically does not exist on the tray, over the weir, in the downcomer, or under the downcomer escape, so it has no hydraulic influence anywhere. With this in mind, the conclusion was not surprising. This 1/5 of the tray liquid which weeps was 1/9 of the overall liquid. 4/9 of the overall liquid flowed across the tray and into the downcomer, with the other 4/9 on the opposite side. The vapor distribution was unaffected. The calculated clear liquid height, total pressure drop, and downcomer backup were all slightly lowered, but there was no hydraulic difference between the weeping side and the non-weeping side.

Final Thoughts

For 2-pass trays, the maldistribution occurs at turndown due to the multiplicity mechanism described in this paper that produces stable but maldistributed hydraulic steady states in which the vapor becomes highly maldistributed in the regions where valves open and close. We expect similar steady state multiplicities at turndown in multi-pass trays (with 3, 4, 6, 8 passes). Worse, these multi-pass trays do not have pressure equalization across the entire tray, as each vapor space is shared by a maximum of two passes. This could cause vapor maldistribution from one tray to propagate several trays upward.

Maldistribution varies the liquid to vapor (L/V) ratio from pass to pass. Lessons from packed tower maldistribution (20-22) extend to multi-pass trays (15, 20). Pinching, and therefore very poor separation, occurs in the low L/V regions, while the excessive L/V ratios in other regions only slightly improve separation. Overall, separation suffers and observed efficiencies diminish, sometimes grossly. Lockett and Billingham (22) show that the efficiency loss depends on the degree of L/V unevenness, as well as the proximity of the pinch.

Tools for reliably predicting multi-pass tray maldistribution are scarce. To our knowledge, no commonly available tray hydraulics rating tool will predict maldistribution in 2-pass trays. It is our hope that the findings and tools in our paper will promote understanding and good design to circumvent the turndown efficiency loss.

Conclusions

This study is the first to report the multiplicity of steady states in 2-pass valve trays at turndown, in the region where some valves open and others are closed. In this region the dry pressure drop becomes constant while the liquid head slowly declines with an increase in vapor load. There are generally three steady states, one with vapor split equally between the passes, while the other two produce stable but maldistributed hydraulic steady states. In the maldistributed steady states, the number of open valves differs from panel to panel, with a corresponding variation in vapor loads. This may cause the vapor to become highly maldistributed between the panels. This multiplicity can lead to unexpectedly reduced turndown in 2-pass valve trays. This multiplicity is found at loadings for which other predictive hydraulic metrics are satisfied.

According to our model, maldistribution due to multiplicity occurs without being induced by maldistribution of adjacent trays or feeds. It is possible whenever the vapor load is such that some valves are open while others are closed. It occurs for pairs of trays which are hydraulically linked by a shared center downcomer.

The distribution ratios (the ratio of the maximum pass L/V to the minimum pass L/V) produced by the maldistributed steady states can exceed 2.5 (a maximum of 1.2 is considered good). An even split of vapor between the two panels (with a distribution ratio of 1), is always one of the steady states. The distribution ratios for the two other maldistributed steady states vary strongly with the vapor rates, approaching unity near the open and closed balance points, and peaking where about half the valves are open.

The multiplicity and distribution ratios are insensitive to weir loads and to weeping. The multiplicity appears to be a vapor-driven phenomenon with little effect on or influence by the liquid.

The maldistribution can be largely alleviated and the distribution ratios reduced by using two valve weights, with 16-gauge and 14-gauge being the most common. For all but the extremely high L/V ratios, this practice would largely reduce the distribution ratios in the regions where valves open and close. However, this practice can also cause a wider range of loadings to experience some maldistribution.

We expect the steady state multiplicity reported here to extend to multi-pass trays. When the distribution ratios are high, we expect a possible significant efficiency loss upon turndown.

References

1. Kister, H. Z., "Distillation Operation", McGraw-Hill, NY, 1990.
2. Kister, H. Z., P. Mathias, D. E. Steinmeyer, W. R. Penney, and J. R. Fair, “Equipment for Distillation, Gas Absorption, Phase Dispersion, and Phase Separation”, Section 14, in R. H. Perry and D. Green “Chemical Engineers’ Handbook”, 8th Ed., 2008.
3. Summers, D. R., L. Spiegel, and E. Kolesnoikov, “Tray Stability at Low Vapor Loads”, Proceedings of “Distillation and Absorption 2010”, Eindhoven University of Technology, The Netherlands, p.611, 2010.
4. Kister, H. Z., R. W. Dionne, W. J. Stupin, and M. Olsson, ”Preventing Maldistribution in Multi-Pass Trays“, Chem. Eng. Prog., p.32, April 2010.
5. Yanagi, T., and M. Sakata, “Performance of a Commercial Scale 14% Hole Area Sieve Tray”, Ind. Eng. Chem. Proc. Des. Dev. 21, p.712, 1982
6. Glitsch, Inc. “Glitsch Ballast Trays A-1, V-1, V-0, Experimental test Data”, Bulletin Nos. 159/160, Dallas, Texas, 1983 (currently obtainable from Koch-Glitsch LP, Wichita, Kansas)
7. Kister, H. Z., “Effect of Design on Tray efficiency in Commercial Towers”, Chem. Eng. Progr., June, p. 39, 2008.
8. Klein, G. F., “Simplified Model Calculates Valve-Tray Pressure Drop”, Chem. Eng., p. 81, May 3, 1982
9. Glitsch, Inc., “Ballast tray Design Manual”, Bulletin 4900, 6th Ed., Dallas, Texas 1993
10. Brierley, R. J. P., P. J. M. Whyman, and J. B. Erskine, “Flow Induced Vibration of Distillation and Absorption Columns”, IChemE. Symp. Ser. 56, p. 2.4/45, 1979.
11. Bolles, W., and J. R. Fair, in J. M. McKetta (Ed.) “Encyclopedia of Chemical Processing and Design”, vol. 16, p. 86, 1982.
12. Colwell, C. J., Ind. Eng. Chem. Proc. Des. Dev. 20 (2), 298, 1981.
13. Buckley, P. S., R. K. Cox, and D. L. Rollins, "Inverse Response in a Distillation Column," Chem. Eng. Prog. 71(6), 1975, p. 83.
14. Summers, D. R., Private Communication, January 3, 2017
15. Bolles, W. “Multipass Flow Distribution and Mass Transfer Efficiency for Distillation Plates”, AIChEJ, 22 (1), 153, 1976
16. Summers, D. R., “Designing Four Pass Trays” Chem. Eng. Prog., p.26, April 2010
17. Bolles, W. L., "Estimating Valve Tray Performance", Chemical Engineering Progress, 72 (9), p. 43, 1976.
18. Colwell, C. J., and J. T. O’Bara, “Estimate Sieve Tray Efficiency in the Weeping Region”, Paper presented at the AIChE Spring Meeting, April 1989.
19. Lockett M. J., S. Banik, Ind. Eng. Chem. Proc. Des. Dev. 25, p.561, 1986.
20. Kister, H. Z., “Distillation Design”, McGraw-Hill, NY, 1992.
21. Pilling, M. “Ensure Proper Design and Operation of Multi-pass Trays”, Chem. Eng. Prog., p.22, June 2005.
22. Lockett, M. J., and Billingham, Trans. IChemE., 80, Part A, p. 373, May 2002; Trans. IChemE., 81, Part A, p. 134, January 2003.

480103 Troubleshooting EB/SM Splitters: How a Maldistribution Analysis Can Help?

Abstract:

Troubleshooting EB/SM Splitters: How a maldistribution analysis can help?

Senthil Krishnamoorthy, Sulzer Chemtech USA Inc.

Lee Ming Yang, Sulzer Singapore Pte. Ltd.

The ethylbenzene (EB) / styrene (SM) splitter (a.k.a. the EB recovery column) is the biggest column in the styrene production process. It can be considered a mega column: column diameters are often larger than 5 m, column heights are taller than 60 m, and 6 beds of structured packing are usually employed.

Generally, packed columns with large diameters are susceptible to large-scale maldistribution. For the EB/SM splitter, the low relative volatility (α-value) of the binary system, and the stringent product specifications, make the column extremely sensitive to maldistribution. Therefore, if an EB/SM splitter is exhibiting poor separation efficiency, a maldistribution evaluation is strongly recommended as part of the troubleshooting evaluation.

This paper will review the various sources of maldistribution susceptibility, which can stem from design, installation, and operational issues. Sometimes, the problem source(s) can only be exactly determined after opening up the column during plant shutdown. As the shutdown time frame is usually limited, there is only a narrow window period to both identify and resolve the source(s) of column underperformance.

Fortunately, it is possible to narrow down the possibilities prior to plant shutdown, via methodologies that encompass process, hydraulic and mechanical considerations. Some important steps include: simulation matchback to operating data, hydraulic evaluations of the column internals, and maldistribution sensitivity analysis of the column. After opening up the column, further methodologies can help to pinpoint the problem source(s). Some important steps include: visual inspection of the column internals, and liquid distributor water tests.

The troubleshooting process will be illustrated via several troubleshooting case studies. They will describe the methodologies used to identify the source(s) of column underperformance, and the subsequent corrective actions to return the column to satisfactory operations. These experiences should help engineers to perform troubleshooting in a structured manner, and allow plant owners to optimize their resources in a troubleshooting activity.

Paper 34a
(Proceedings 479206)

Vacuum Distillation:

Wash Zone Configuration

Andrew W. Sloley
Advisian (WorleyParsons Group)
Energy Center II
575 N Dairy Ashford St
Houston, TX 77079

Prepared for Presentation at the
American Institute of Chemical Engineers
2017 Spring Meeting
San Antonio, TX March 26-30, 2017

UNPUBLISHED

1. Summary

Refinery vacuum units include a deentrainment, or wash, zone for removal of heavy material from desired products. Wash zone reliability problems often shut vacuum units down, causing significant losses. Factors that strongly influence wash zone reliability include vapor and liquid distribution, upstream flow regimes, and wash zone efficiency and pressure drop. For fuels vacuum units, the wash zone should have the minimum amount of packing consistent with acceptable deentrainment. The minimum wash bed has lower capacity and higher reliability.

One major factor that must be evaluated on a case-by-case basis is phase separation in the transfer line. Even small amounts of phase separation will feed superheated vapor to the wash zone. Superheated vapor dramatically raises the amount of wash oil required for wetting the wash bed. Properly installed and operated, some vacuum towers have demonstrated trouble-free performance for 20+ years. Others, with improper understanding of the operation, suffer from one to two year run lengths, dramatically reducing refinery profits.

2. Introduction

This discussion has the intent of covering process considerations for wash zone operations. However, equipment constraints and process requirements interact. Process conditions must remain within equipment operating constraints for reliable and effective operation. Equipment constraints will be mentioned as required to explain the process limits and their importance. Previous work has focused on equipment constraints and includes more detail on them [1].

3. Operating Objectives

Refinery main fractionators include a wash section immediately above a two-phase feed. These towers have very high inlet velocities and high kinetic energy in the feed. The high inlet velocity and energy dissipation in the feed zone of the tower easily creates small liquid droplets. The droplets entrain into the rising vapor. Without a wash zone, the liquid droplets end up in the products above the feed.

The wash zone is a deentrainment device to remove the liquid droplets. The vacuum distillation tower works in a severe operating environment. Figure 1 illustrates a typical refinery vacuum tower and its wash zone. The wash zone is a packed bed above the tower feed.

The wash section has the following major objectives:

1. Deentrain liquid from the rising vapors
2. Maximize yields by maintaining low pressure drop and low wash liquid rate
3. Maximize reliability
4. Fractionate volatile contaminants (metal compounds)

These objectives compete with each other. Wash zones that meet some objectives tend to have problems with the other objectives.

4. Terminology

Figure 2 shows the relevant components of the wash bed, and the external and internal streams. The definitions of the streams for this discussion are:

Table 1 Wash Zone Streams

Stream	Definition
Wash (or wash oil)	Liquid placed on top of the wash bed to irrigate the bed to prevent coking.
Overflash	Distillate liquid leaving the bottom of the wash bed. This is material that was vaporized in the flash zone.
Entrainment	Liquid droplets moved up the tower by high vapor velocity. This is material that was not vaporized in the flash zone.
Slop wax [1]	The combination of entrainment and overflash that is collected on the tray below the wash bed. Slop wax = overflash + entrainment
Condensate	Liquid condensed on the bottom of the collector tray because the overflash is colder than the feed vapor. This stream is shown for completeness. In most units it has a very low flow rate.

If the entrainment rate is high, most of the liquid gathered by the slop wax collector is entrainment, not a recoverable distillate.

1 The term slop wax comes from lubricants vacuum units. The bottoms lubricant cut had a maximum allowed viscosity or viscosity index. The bottoms of the tower went to processing for bright stock (cylinder oil) recovery with a minimum viscosity specification. The slop wax was a drag stream between the two products that removed material with too high a viscosity for the lubricant draw but too low a viscosity for the bright stock.

5. Fractionating Versus Deentraining Wash Zones

Most refinery vacuum units are for fuels production. The vacuum tower products go to downstream processing for producing fuels. Many fewer vacuum units produce lubricant base-stocks. Lubricant production is only a small fraction of total vacuum unit capacity (<10%). This discussion focuses on fuels vacuum units. Nevertheless, many of the concepts apply to lubricants vacuum units.

Fuels vacuum units may have two purification objectives for the wash zone:

1. Remove mechanical entrainment of small liquid droplets from the vapor, or
2. Fractionate volatile materials to improve the lower product (HVGO) quality.

Mechanical entrainment puts vacuum residue in the distillate product. This can cause large problems with asphaltenes, metals, and carbon residue qualities of the HVGO.

The most common volatile contaminants of interest are vanadium and nickel porphyrins. High porphyrin levels are most common in heavy crudes (API gravity <= ~21.9°). Lighter crudes tend to have much lower porphyrin concentrations. Crudes well known for volatile metal compounds include Bachaquero, Leona, Isthmus, and Maya, as well as many others.

When the wash bed objective includes fractionation, fractionation requirements set the packed bed efficiency and liquid requirements. This efficiency and liquid rate tend to be much higher than the requirements for a wash bed intended to meet a deentrainment specification.

This discussion focuses on wash beds for deentraining services.

6. Conventional Configuration

Figure 3 shows a conventional configuration for a wash zone. Generic elements include:

1. Wash oil distributor to put liquid on the top of the wash zone packing.
2. Wash zone packing to act as a deentrainment device.
3. Collector under the wash zone to collect overflash and distribute vapor entering the wash zone.

Other elements in Figure 3 may vary between different conventional vacuum units but have little effect on the overall analysis of wash zone operation. In this case:

1. Overflash is recycled to the tower boot with an external pump. Other options include recycling the overflash upstream and allowing internal overflow of the overflash. Overflash disposition has a negligible effect on the current analysis.
2. The vacuum tower has no stripping trays. Stripping trays use steam to increase overall product yield. The presence of stripping trays has only a minor effect on the wash zone analysis here.

7. Conventional Operations

7.1. Ideal Wash Zone

Figure 4 compares an ideal flash zone to an achievable flash zone. Actual flash zone performance never achieves perfect deentrainment. Close-to-perfect deentrainment requires low vapor velocity to allow droplet settling and very few small liquid droplets (< 100 microns). Capital costs keep flash zone sizes small. Some entrainment always occurs.

A deentrainment or wash section above the feed removes the entrainment. To prevent coking, the deentrainment equipment requires net liquid flow at all points. The distillate liquid entering the wash zone is wash oil. The liquid leaving the wash zone is material that had to be flashed "over" the requirement to meet the yields observed. Hence the name overflash for distillate liquid leaving the wash zone.

With deentrainment as the primary objective, the ideal wash zone would have:

1. Minimum fractionation efficiency (zero if possible).
2. Minimum pressure drop (zero if possible).

No physical device can attain perfect performance. The comparison basis used here has been defined as:

1. One theoretical stage flash zone.
2. Minimum pressure drop (2 mmHg) across the slop wax collector tray.
3. Minimum pressure drop (1 mmHg) across the wash zone.

Various curves are presented to show how yield and unit capacity change with wash zone operation. Yield is quantified by cutpoint, the atmospheric equivalent temperature that would have the same distillate yield as the operating unit as on the ASTM D2892 [2] test. High cutpoints equal more distillate yield from the same feed. Capacity factors are relative unit feed rate. High capacity factors equal a higher feed rate to the vacuum unit.

7.2. Operating Limits

Minimum wash oil rates vary as a complex function of:

- Tower diameter
- Fractionation across the wash zone
- Mass flow rate
- Pressure drop across the wash zone
- Choice of mechanical equipment used

The objective of the wash oil is to create a targeted overflash rate. The liquid rate leaving the bottom of the bed strongly correlates with coking rates and reliability. Due to energy balance and fractionation effects, the wash oil rate will always be more than the overflash rate. The wash bed dries out from the top to the bottom [3]. The wash bed has the highest liquid rate at the top of the bed and the lowest liquid rate at the bottom of the bed. The dry-out ratio is the ratio of the liquid volume at the top of the bed to the liquid volume at the bottom of the bed:

$$dry-out\ ratio = \frac{liquid\ volumetric\ rate\ at\ the\ top\ of\ the\ wash\ bed}{liquid\ volumetric\ rate\ at\ the\ bottom\ of\ the\ wash\ bed}$$

The dry-out ratio is often quoted using standard liquid volume rates. However, here the actual flowing liquid volumes are used.

If the overflash rate is too high, yield drops as excess overflash increases the bottoms rate. If the overflash rate is too low, the wash bed can coke and the unit is forced to shut down. Both problems have a high economic cost.

This creates the fundamental operating challenge for vacuum tower wash zones:

1. The economic cost of using too high a wash rate is high.
2. The economic cost of using too low a wash rate is high.
3. Operating conditions are severe.

Overflash can be defined as a fraction of feed to the unit or a liquid load based on wash bed cross-sectional area. Older references often take the first approach, using a defined percent of unit feed. The most quoted range of overflash rates are based on Watkins [4, p. 110] and use 1 to 2 % overflash for conventional crudes and up to 4% overflash for high-metal content crudes. It is defined as:

$$overflash\ percent = 100\frac{overflash\ rate, std\ volume}{feed\ rate, std\ volume}$$

Many authors have used 2% as the starting point for unit optimization studies. This is consistent with the design curve included in Watkins.

The second approach uses overflash rates defined by actual flowing volumes and tower diameter. Various authors have suggested minimum overflash loads varying from 0.125 to 0.180 gpm/ft^2 of cross section area at the bottom of the wash bed (5.09 to 7.33 lpm/m^2) based on volume at conditions. The examples used here are based of 0.150 gpm/ft^2 (6.11 lpm/m^2) overflash load. The examples use a maximum vapor rate equivalent to a C-factor of 0.45 [1].

Figure 5 plots overflash percent versus overflash rate. The plot shows that a liquid load of 0.15 gpm/ft^2 (6.11 lpm/m^2) corresponds to 4.3% overflash for the example case used here. Thus, operation at the minimum recommended liquid load has 2.15 times the conventional recommended overflash percent. Clearly, the recommendations are highly inconsistent.

Operating pressure is the likely source of the difference. The worked example in Watkins uses 2% overflash and 42 mmHg flash zone pressure. Many modern vacuum towers run down to 20 mmHg (or lower) flash zone pressure. At low pressures, vacuum tower diameter increases. The larger vacuum tower needs a higher volume of overflash to get the minimum required rate at the bottom of the bed.

Extensive industrial experience has shown that targeting a minimum load (gpm/ft^2) is a fundamentally more sound approach to unit design and operation. Overflash percents should not be used.

Figure 6 and Figure 7 show operating changes versus overflash rate. As overflash rates increase, capacity decreases and yield drops.

7.3. Fractionation

The more fractionation in the wash bed,

1. The larger the composition gradient across the bed.
2. The larger the temperature gradient across the bed.
3. The larger the dry-out ratio from thermal effects.
4. The higher the revaporization rate and the larger the tower diameter for the same capacity or the lower the capacity for the same diameter.
5. The higher the wash oil rate for the same overflash rate.
6. The lower the HVGO draw temperature for the same yield (minor effect).

In general, an ideal wash bed would deentrain but have zero fractionation efficiency. Figure 8 illustrates changes in dry-out ratio, tower capacity, and relative wash oil rate with changes in the number of theoretical stages in the wash zone. Higher fractionation efficiency increases dry-out ratio, reduces unit capacity, and increases the needed wash oil rate. Figure 9 shows cutpoint (yield) versus wash zone stages. The more wash zone stages, the lower the cutpoint.

The largest change is moving from zero theoretical stages to one theoretical stage in the wash zone. While more stages do increase the wash oil rate for the same overflash, the increase is minor.

7.4. Pressure Drop

Pressure drop with no liquid contacting the rising vapor increases the amount of vapor superheat at the top of the bed. Higher vapor superheat increases the wash oil rate required. Some of the wash oil must vaporize to get the vapor down to its dew point.

Higher pressure drop has two boundary conditions:

1. Higher pressure drop with a fixed flash zone pressure causing the pressure at the top of the wash bed to be lower.
2. Higher pressure drop with a rising flash zone pressure and a fixed pressure at the top of the wash bed.

The first case, higher pressure drop with a fixed flash zone pressure, approximates a design case with a target yield defined by flash zone pressure and temperature. Figure 10 and Figure 11 show the dry-out ratios, relative capacity, and cutpoint for varying wash zone pressure drops for the first case.

The second case, higher pressure drop with a rising flash zone pressure, approximates an operating case limited by vacuum system capability. Figure 12 and Figure 13 show the dry-out ratios, relative capacity, and cutpoint for varying wash zone pressure drops for the second case.

The effect of wash zone pressure drop on dry-out ratio is small. The design case with a lower pressure at the top of the wash zone has a large impact on unit capacity (or size). The hydraulic limit in the wash zone is nearly always at the top of the bed. Therefore, reducing the pressure at the top of the bed creates a large capacity drop.

Figure 12 shows increasing capacity with pressure drop across the wash bed. The calculations include changes in transfer line operation with changing flash zone pressure. This capacity increase is due to changes in the transfer line. Section 7.8 includes more detail on transfer line effects.

7.5. Chimney Tray

Figure 1 and the other figures all show vacuum towers with a collector tray underneath the wash zone packing. Some vacuum towers do not have a collector tray. Without the collector tray, yields tend to be lower and deentrainment performance worse.

The collector tray:

1. Prevents the colder slop wax from quenching the flash zone vapors. This increases overall yield.
2. Reduces slop wax re-entrainment. This improves product quality.
3. Improves vapor distribution to the packing. This reduces local areas with high vapor rates and helps reduce localized coking rates. The result is more reliable operation.
4. Collects slop wax for possible metering and sampling. This gives the plant potential tools for improved optimization and increased yields.

The collector tray should be installed. However, this comes with more than an equipment cost. The collector tray creates pressure drop. In general, the higher the pressure drop, the better the vapor distribution above the tray. The collector tray pressure drop can be calculated using an orifice equation. These calculations are usually accurate for the HVGO and LVGO distributors above the wash section. However, for the slop wax collector, measured pressure drops are routinely twice the calculated value.

The higher measured value is caused by vapor flow maldistribution. Computational fluid dynamic analysis of tower entry and collector systems has shown complex flow patterns. These patterns often have severe local variations in velocity and direction [5]. If the chimney tray pressure drop is low and the feed entry poor, parts of the chimney tray may even have vapor flowing down instead of up through the chimneys. A better feed entry device and a higher collector tray pressure drop can dramatically improve vapor distribution to the bed above.

With a good feed entry device, the target pressure drop for the collector tray should be 1 mmHg. In service, the expected pressure drop is closer to 2 mmHg. The author has seen units with as much as 10 mmHg pressure drop across the collector tray.

Pressure drop across the collector tray has nearly identical consequences as extra pressure drop across the wash zone packing.

7.6. Loading (C-factor)

Vapor-to-liquid mass ratio affects the dry-out rate. Section 7.2 discusses the reduced dry-out ratios with higher liquid rates. This is because the higher liquid rate has a higher L/V ratio in the wash section. Less of the wash oil vaporizes to absorb the energy from the cooling vapor.

Lower vapor rates have the same results. Low vapor rates give low C-factors in the packing. Figure 14 and Figure 15 show the dry-out rate, capacity factor, and cutpoint for a constant 0.150 gpm/ft^2 (6.11 lpm/m^2) overflash rate at different C-factors. Lower C-factor reduces the dry-out ratio but also reduces the yield.

Figure 16 shows how the column area, wash oil rate, and overflash rate change as the C-factor changes for a constant feed rate. Higher C-factors are equivalent to higher mass flow rates. The change in mass flow rate is the underlying cause of the change in dry-out ratio. While the higher C-factor produces a higher dry-out ratio, it also requires a smaller area. The change in area is more important than the change in dry-out ratio.

This is also the reason why the higher operating pressure mentioned in Section 7.2 gives a lower overflash percent. The change in wash oil rate is the product of two factors:

1. Higher operating pressure increases the density and increases the max rate per unit area, increasing the dry-out ratio.
2. Higher operating pressure decreases the column cross section area, decreasing the irrigation area.

$$Wash\,Oil\,Volumetric\,Rate = Overflash\,Load \times Dry-out\,Ratio \times Cross\,Section\,Area$$

Dry-out ratio and cross section area change in opposite directions when pressure changes. The change in cross section area is more important than the change in dry-out ratio. Hence, as pressure increases, the total volume flow of wash oil drops.

7.7. Entrainment

Entrainment can be the source of liquid on the wash bed. In some units the overflash rate is approximately zero (< 0.02 gpm/ft^2, 0.81 lpm/m^2). In spite of this, multiple-year runs without coking the bed were achieved.

With no overflash, the slop wax is 100% entrainment. In these units, the slop wax rate was measured and was found to be very high (over 10% of the total residue rate from the bottoms). The crude was clean, light, and thermally stable.

Under these conditions, entrainment irrigated the bottom of the wash bed. The small amount of overflash used irrigated the top part of the bed. Entrainment reached far enough into the wash bed that it overlapped the region where overflash was present. The entire bed was irrigated and did not coke up. Figure 17 illustrates this operation. Intentionally operating in this mode is not recommended. The probability of coking the bed is high. Only small changes in either wash rate or entrainment rate can lead to coking or severely off-specification HVGO product.

7.8. Transfer Line Phase Separation

In the early 1990s the author and others reviewed several severe coking incidents where everything indicated that the wash bed should have been sufficiently irrigated. However, plant data was showing very little overflash. In these cases, the transfer line allowed for phase separation between the vapor and the liquid.

With phase separation, the depressurizing vapor becomes superheated. The superheat is not available to vaporize more liquid in the transfer line. The limited surface area between the vapor and liquid limits the amount of heat transfer. Figure 18 illustrates this operation.

The basic technique of accounting for phase separation in the transfer line has been referred to as the non-equilibrium model for transfer lines [6]. The author uses a derivative of the basic approach and prefers to call this the phase-slip model because phase-slip avoids confusion about the topic. The term equilibrium model is already in wide use in the discussion of equilibrium-stage distillation models versus rate-based distillation models.

As well as reducing yield, the superheated vapor entering the vacuum tower increases the dry-out ratio. The superheated vapor vaporizes some of the overflash as the vapor-liquid mixture reaches equilibrium. Figure 19 and Figure 20 shows the effect of phase slip.

Phase slip index (φ) has been defined as:

$$\varphi = \textit{phase slip index} = 100\,\frac{\textit{pressure drop total} - \textit{pressure drop before phase separation}}{\textit{pressure drop total}}$$

Transfer line phase separation is one of the most difficult factors to evaluate when designing a unit. With data, the results of phase separation are clear. Getting the data, however, can be very challenging. The base configuration used for all the comparisons had a transfer line with a φ=29. Values of 10 to 30 are typical for many operating units.

Based on data, unit operations can cover the entire range. Some units have achieved φ of 0. Others have demonstrated yields and required wash oil rates consistent with φ of over 90. Phase separation is the single largest factor in uncertainty in understanding wash zone operation on many units.

Transfer line phase pressure drop and phase separation are highly dependent on transfer line geometry, overall pressure profile, and operating conditions. Nevertheless, factors that make transfer line separation more likely include:

1. Low heater outlet pressure.
2. Low vacuum tower pressure.
3. Large diameter transfer lines.
4. Reduced rate operation.

7.9. Residence Time

Time and temperature combine to create coke. Keeping the oil hot for a longer period of time increases the risk of coking the wash bed. Residence time is not directly included in ideal calculations, but should never be overlooked. Higher irrigation rates (higher wash oil rates) reduce the residence time in the wash bed.

7.10. Interactions

The author has covered major process issues in the wash zone that affect performance. Each has been covered independently. In actual units, the factors combine. Usually, higher efficiency beds:

1. Have higher pressure drop (use more, or more efficient, packing).
2. Have more liquid hold-up because of the higher pressure drop resulting in longer residence time.
3. Use smaller packing sizes and aggressive surface treatments (to increase efficiency) so they are more likely to trap coke in the packing.
4. Have more stages, so they are more subject to local dry out problems due to limits in vapor and liquid distribution.

Combined with transfer line phase separation and problems with feed characterization [7] and molecular weight prediction [8], dry-out ratios can be high. Field data have shown some units have dry-out ratios as high as 15:1.

8. Hot-Wash Configuration versus Cold-Wash Configuration

Hot wash units use the direct liquid drawn from the collector above the wash section as the wash oil. This is normally an HVGO stream. The hot wash is a bubble point liquid on the collector tray and remains a bubble point liquid when it is returned to the top of the wash packing.

Cold wash units use a liquid that comes from downstream in the HVGO cooling circuit as the wash oil. The cold was stream is a subcooled liquid when returned to the top of the wash packing.

Figure 21 and Figure 22 illustrate the hot wash and cold wash configurations.

As long as the liquid rate leaving the bottom of the wash bed is the same, the amount of liquid on the top of the bed is not particularly relevant to meeting process objectives. However, a number of practical concerns favor hot wash.

The higher liquid rate to the top of the bed in hot wash allows for larger openings in the liquid distributor. This makes the distributor less likely to plug or foul. Additionally, subcooled liquids sprayed into a rising vapor cause some of the vapor to condense. The condensing vapor can pull the liquid cones in and collapse part of the cone. Spray cone collapse can lead to portions of the top of the bed having less liquid than expected. Liquid rates in the wash packing are so low that even small deviations in liquid distribution can cause accelerated coking in the wash bed. Hot wash should be used instead of cold wash.

9. Wash Oil Distributors

Two types of liquid distributors are used in wash oil services:

1. Spray header distributors.
2. Trough distributors.

Mechanical and service details of these have been previously covered [1]. In addition to mechanical considerations, process performance of the wash oil distributors affects wash zone performance.

Data currently available do not clearly support one type over the other. Both can be used. Both have benefits and costs. Overall, the spray distributor is usually cheaper to purchase, and easier and cheaper to install. Unless a specific reason is present, the spray distributor is the recommended choice.

Table 2 summarizes the distributor characteristics. Major factors include:

1. Entrainment of distributor liquid before it reaches the packing.
2. Distribution quality and fractionation performance of the distributor.
3. Heat-transfer performance of the distributor.
4. Size of distribution holes.

Table 2 Distributor Characteristics

	Spray Distributor	Trough Distributor
Entrainment of wash liquid	Small droplets created at spray nozzles. Liquid splashing when entering packing.	Vapor channeling due to high vapor velocity when it passes between troughs can entrain liquid from the packing.
Distribution Quality	Low quality distribution reduces fractionation effectiveness (desired) but can lead to low liquid in some areas that might accelerate coking.	Smooth distribution of liquid increases fractionation effectiveness. This is undesired unless fractionation for metals control or viscosity control is required.
Heat-Transfer Performance	Sprays are excellent heat transfer devices. This is undesired.	No heat transfer above the packing. But the good liquid distribution increases heat transfer in the packing.
Size of Holes	Holes in spray nozzles slightly larger.	Very small holes for most services. This is a potential problem even with deflector plates used to increase hole sizes.
Other	Lower cost and relatively simple installation.	Higher cost. Installation can be labor intensive in large-diameter columns. More likely to be damaged by pressure surges.

10. Spray Chambers

In a column without internals, there is nothing for coke to attach to. Spray chambers use one to three rows of spray nozzles. The wash liquid contacts the rising vapor. The small rising droplets combine with larger droplets coming from the spray nozzle. Liquid drops agglomerate by direct impact and by Brownian motion. Delayed coker main fractionators routinely use spray chambers. The delayed coking service is an extremely severe coking service.

Figure 23 illustrates a vacuum tower with a spray chamber. When trays were still used in vacuum towers, having a spray chamber could significantly reduce flash zone pressure. Additionally, trays have a poor record of rapid coking and relatively high entrainment. Trayed towers often used wire-mesh pads [9]–[15]. The mesh pads then created their own problems with coking, high-pressure drop, and collapse. With annual turnarounds common in the 1950s, plants could tolerate the mesh pads coking. With current multiple-year runs, mesh pads have no place around the wash section of vacuum towers.

In spite of the problems with trays and mesh pads, they had better deentrainment performance than spray chambers. To get adequate deentrainment, spray chambers need to operate with an excellent inlet feed distributor and low C-factors. An upper limit on C-factor of 0.20 should be used with spray chambers. A more prudent limit would be a C-factor of 0.18 or below. A spray chamber is really a separator drum. For comparison, most separators without internals are designed for equivalent C-factors of 0.15 to 0.175.

11. Recycle Wash Systems

Some units have recycle wash systems to increase the amount of wash oil while reducing the overflash rate. These systems take some of the slop wax and add it back to the wash oil. Figure 24 shows a one-stage recycle wash system. Figure 25 shows a two-stage recycle wash system. The vacuum tower has two separate wash beds. The upper wash bed uses HVGO as the wash oil, the same as a conventional system. The lower bed uses 100% slop wax as the wash oil. The wash oil from the upper bed falls down into the lower bed. This creates a purge for the system.

The refiner must keep in mind that the recycle contains entrainment. Potential problems with this include:

1. Increased residence time at a high temperature for the recycled entrainment – possibly accelerating coking.
2. Fouling and plugging of the packing or distributors with solids (scale, clay, rust, iron sulfide) contained in the entrainment.
3. Impurities in the HVGO from entrainment of the wash oil return.

Recycle systems have few benefits. The situations where they work well include units running relatively clean and thermally stable crudes. With very few exceptions, the recycle wash systems analyzed by the author could have been replaced with better or lower cost options.

12. Case Study 1

A refiner converted a trayed vacuum tower into a packed vacuum tower. Figure 26 illustrates the history of modifications made. The trayed tower had three trays in the wash zone. The original configuration is shown in detail '0' in Figure 26. The tray spacing was:

Table 3 Vacuum Tower Tray Spacing

Tray	Elevation	Spacing	Spacing
3	31'-0"		
		5'-6"	1.676 m
2	25'-6"		
		5'-6"	1.676 m
1	20'-0"		
		7'-6"	2.286 m
Feed	11'-6"		

The large tray spacing between the second and first tray was intended to reduce the amount of entrainment. Deentrainment improvements with extra tray spacing dramatically decrease once tray spacing exceeds 4 ft (1.219 m). The extra spacing probably had minimal effect on tower performance.

12.1. Revamp 1: Replace Trays With Packing

The refiner was attracted by the benefits of lower pressure drop and higher capacity of structured packing. Trays 1-3 were removed and replaced with a combined packing-grid bed. Total bed depth was approximately 11' (3.35 m). The bottom 5' (1.52 m) was grid and the upper 6' (1.83 m) was structured packing. A spray header was installed for liquid distribution. Based on expected packing efficiency the bed was estimated to have four ideal stages of separation. The estimate of four theoretical stages assumed that the spray header would perform as a high-quality liquid distributor.

The installation was mechanically convenient. The lower elevation was the existing Tray 1 support. No slop wax collector was installed. Figure 26 shows these modifications in detail '1'.

The wash oil rate was set based upon a 2:1 dry-out ratio. This was a common basis used for older vacuum towers with trays. No data were available from the unit as overflash flowed internally into the bottoms.

Performance did not meet the expected goals. Entrainment became severe at a C-factor of 0.38 compared to the predicted capacity limit of 0.5. Additionally, after approximately 16 months of operation, bed pressure drop started to rise. Entrainment rapidly increased as the pressure drop increased. The unit was shut down after 24 months.

12.2. Revamp 2: Add a Slop Wax Collector and Slop Wax Recycle

The possible causes of the rapid coking and lower than expected capacity were a matter of contentious debate. No data were available on actual overflash rates. The decision was made to install a slop wax collector, increase the wash oil rate, and recycle the slop wax to the heater feed. Figure 26 detail '2' shows these modifications.

The reasoning included:

1. A higher wash oil rate would reduce the coking rate.

2. Recycling the slop wax would allow for recovery of distillate in the slop wax. Therefore, the higher overflash rate would not reduce yields.
3. The collector and pump would allow for sampling the slop wax. This would give the plant better data for optimization.
4. The chimney tray would reduce re-entrainment of liquid by separating the slop wax from the vapor in the flash zone.

That the chimney tray might improve wash zone performance by improving vapor distribution was not considered. Later analysis would show that the improved vapor distribution was important.

The modifications did improve the operation. Capacity rose to a C-factor of 0.42 at the same level of entrainment seen after the 1st set of modifications. While improved, this was still not the promised value of 0.5.

Samples revealed nearly zero overflash in the slop wax. Both an asphaltene balance and a sodium balance showed that the slop wax collected was over 95% entrainment. This was true with double the wash oil rate used in the previous run.

If the slop wax is nearly 100% residue, then it contains no distillate to recover by recycling it to the furnace feed. Operating events supported this analysis. The slop wax pump had an operating upset and the pump had to be removed from service. Unit yields did not change when the pump was out of service.

Based on matching a measured overflash in the slop wax, the wash bed was estimated to have a 15:1 dry-out ratio. This seemed very high. Referring to Figure 8, at three theoretical stages the dry-out ratio is less than 7:1 and the incremental change per stage is already very low. It was difficult to understand how 15:1 could occur. The most popular suggestion was that small errors in laboratory technique were causing the dry-out ratio to be over-estimated.

At the start, gas oil quality was very good. The gas oil had a 2 color and no significant asphaltene content. Quality stayed at this level until the pressure drop started to increase.

No significant improvement in run length was seen. After approximately 16 months, pressure drop started to rise again.

12.3. Revamp 3: Reduce Wash Bed Height

At this point the system was clearly not understood and the path forward was not clear. Several other units were suffering from similar wash zone problems at this time. Different strategies were being used at different plants to understand the problem.

Since product quality had been so good at the start of the run, the idea was that less packing was needed. The perceived benefit was that less packing would reduce residence time in the bed and reduce the coking rate.

Figure 26 detail '3' shows the modification made. Wash bed height was reduced from 11' (3.35 m) to 5' (1.8 m). The bed included roughly 40" (1 m) of grid and 20" (0.5 m) of structured packing.

This change resulted in a lower initial pressure drop (less packing), and HVGO quality still met downstream processing requirements. The quality was better than that seen after the first revamp, but not as good as after the second revamp. This was as expected.

Run length did increase. The unit ran for approximately 20 months before pressure drop started to rise. The unit reached 36 months before needing to be shut down for packing replacement. The extra run-length came from a combination of three factors:

1. A lower starting pressure drop (less packing depth).
2. A longer duration before pressure drop started to rise (20 months versus 16 months).

3. A slower rate of increase (due to lower coking rate from lower residence time).

Sampling of the slop wax showed a 7:1 dry-out ratio. This seemed high for the one theoretical stage expected from the reduced packing height.

Modeling with calculations forced to match the dry-out ratio showed that the maximum operation had a 0.45 C-factor, not a C-factor of 0.42. The higher dry-out ratio creates a higher calculated C-factor at the top of the bed.

Experience from other units supported the 0.45 C-factor limit for acceptable entrainment. The original 0.5 C-factor estimate for this crude was over-optimistic.

During this run, data from other units was showing that phase separation in the transfer line was becoming important. Due to the cost of modifying the transfer line, no modifications were proposed.

12.4. Revamp 4: Relocate Wash Oil Distributor

The packing still had insufficient wash oil. Calculations with an assumed phase slip of 50 (φ) matched a 7:1 dry-out ratio. Modifications were planned to increase the wash oil rate to the 7:1 dry out ratio. Spray header modifications were implemented in the 4th revamp.

During this work, it was noticed that the inside of the tower showed evidence of significant liquid spray on the vessel wall. The previous revamp (#3) had reduced the packing height, but not relocated the spray header. A quick piping modification allowed the spray header to be dropped to the correct elevation above the packing.

Plant data after these modifications showed the expected dry-out ratio to match the prediction. Gas oil quality was excellent and the full desired run length of 5 years was achieved.

12.5. Revamp 5: Update Feed Entry

To reach the maximum capacity, a new feed entry device was proposed for the fifth shutdown. The idea was that if less entrainment was carried into the wash bed, vapor rates could be pushed to a higher C-factor. Some entrainment would occur, but product quality would remain acceptable.

After installation of the new feed entry device, rates could be pushed as high as a C-factor of 0.47. However, minor operating changes would cause the tower to flood rapidly. The decision was made to limit feed rates to the 0.45 C-factor.

The new feed entry device improved the HVGO quality slightly and also reduced slop wax rates a small amount.

12.6. Overview of Case 1

Modifications unfolded over a number of years. At the start, the potential for wash zone coking was not well understood. The unplanned factors that contributed to wash zone coking were phase separation in the transfer line and high efficiency in the packed bed. The phase separation required higher wash oil rates than originally planned. The wash bed was reduced in height to reduce the number of theoretical stages. After modifications were made, run length met the original targets. Further developments that improved operations included adding an improved feed entry device.

13. Case Study 2

A refiner upgraded a vacuum tower wash zone from trays and wire-mesh mist eliminators to a packed bed using a 3rd generation random packing. The packing was 6' (1.83 m) of rings. The planned operation was to run with 0.15 gpm/ft^2 (6.11 lpm/m^2) of overflash. The slop wax was collected and recycled to the furnace feed. The overflash pumps always had reliability problems and rarely remained in service for a complete 4-year run of the vacuum tower.

The packing coked during every run. Plant data showed high pressure drops in the packed bed starting approximately 6-8 months after the run started. The bed was always coked when examined at the end of a run. The bed was replaced in-kind in subsequent runs. In spite of the coking, the wash oil rate was not increased.

Even though the bed coked, the rings had superior performance (lower pressure drop and higher capacity) than the wash trays and mist eliminator pads that were removed.

The current work was intended to upgrade the vacuum tower from the 3rd generation random packing to a mix of structured packing and grid. The wash bed was upgraded during modifications that added a diesel product to the vacuum tower. The new wash bed used ~4 ft (1.22 m) of a mix of grid and structured packing. A new collector tray, slop wax draw nozzle, and slop wax pumps were installed. The slop wax pumps were converted to use variable frequency drives to control flow rate instead of using a control valve. This improved pump reliability by reducing average suction specific speed of the pump [16]. Wash oil rates were re-set based on an approximate estimate of a φ of 50 and a targeted overflash rate of 0.15 gpm/ft^2 (6.11 lpm/m^2).

After modifications, laboratory testing of the overflash showed rates slightly above 0.15 gpm/ft^2 (6.11 lpm/m^2). Over a period of weeks the rate was optimized. Assuming a φ of 30 matched the plant data.

Operation has been consistent and no evidence of coking has been seen from the operating data.

14. Conclusions

Wash zones remain areas where refiners can make, or lose, large profits. Reliable operation allows for high capacity, yield, and quality. With sufficient wash oil, packed beds are very reliable in wash zone service. Factors influencing the right wash oil rate remain unknown to many operators. This review covered a partial list of the factors including:

- Efficiency
- Pressure drop
- Operating rate
- Transfer line phase separation
- Proper liquid and vapor distribution

Properly installed and operated some vacuum towers have demonstrated trouble-free performance for 20+ years. Others, with improper understanding of the operation, suffer from one to two year run lengths, dramatically reducing refinery profits.

15. References

[1] A. W. Sloley, "Refinery Vacuum Towers: State of the Art," presented at the AIChE National Meeting, San Francisco, CA, 2013.

[2] ASTM, "D2892-05, Standard Test Method for Distillation of Crude Petroleum (15-Theoretical Plate Column)," West Conshohocken, PA, 2005.

[3] A. W. Sloley, "Atmospheric Distillation Process: Fundamental Concepts," presented at the AIChE Spring Meeting, New Orleans, LA, 2014.

[4] R. N. Watkins, *Petroleum Refinery Distillation*, 2nd ed. Houston, TX: Gulf Publishing Company, 1969.

[5] K. Kolmetz, W. Kiong, K. Ng, Y. T. Senthil, A. W. Sloley, and T. M. Zygula, "Optimize process operation with new vacuum distillation methods," *Hyd Proc*, vol. 84, no. 5, pp. 77–85, May 2005.
[6] S. W. Golden, V. Shah, and J. P. Kovatch, "Improved Flowsheet Topology for Vacuum Distillation Simulations," presented at the Canadian Chemical Engineering Conference, Calgary, ALB, Canada, 1994.
[7] D. C. Villalanti, J. C. Raia, and J. B. Maynard, "High-Temperature Simulated Distillation Applications in Petroleum Characterization," in *Encyclopedia of Analytical Chemistry*, Chichester, UK: John Wiley & Sons, 2000, pp. 6726–6741.
[8] K. H. Altgelt and M. M. Boduszynski, *Composition and Analysis of Heavy Petroleum Fractions*. New Orleans, LA: Marcel Dekker, Inc., 1994.
[9] A. H. Orr, "Use of Wire-Mesh Sections in Asphalt Entrainment Control in Vacuum Pipe Stills, Part 1... Case History - Cosden Petroleum Corp.," *OGJ*, pp. 84–85, Nov. 1953.
[10] D. H. Stormon, "Use of Wire-Mesh Sections in Asphalt Entrainment Control in Vacuum Pipe Stills, Part 1... Case History - General Petroleum Corp., Torrance, Calif.," *OGJ*, pp. 87, 121–122, Nov. 1953.
[11] H. S. Fourer, "Use of Wire-Mesh Sections in Asphalt Entrainment Control in Vacuum Pipe Stills, Part 1... Case History - Sincalir Refining Co., Marcus Hook, Pa.," *OGJ*, pp. 86–87, Nov. 1953.
[12] J. S. Corlew, "Use of Wire-Mesh Sections in Asphalt Entrainment Control in Vacuum Pipe Stills, Part 1... Case History - Sinclair Refining Co., Sinclair, Wyo.," *OGJ*, pp. 82–84, Nov. 1953.
[13] J. W. Russell, "Use of Wire-Mesh Sections in Asphalt Entrainment Control in Vacuum Pipe Stills, Part 2... Case History - Ohio Oil Co., Robinson, Ill.," *OGJ*, pp. 95–96, Nov. 1953.
[14] C. O. Anderson, "Use of Wire-Mesh Sections in Asphalt Entrainment Control in Vacuum Pipe Stills, Part 2... Case History -Rock Island Refining Co. Indianapolis," *OGJ*, p. 99, Nov. 1953.
[15] J. H. Fisher and L. R. Bunger, "Use of Wire-Mesh Sections in Asphalt Entrainment Control in Vacuum Pipe Stills, Part 2... Case History - Sinclair Refining Co., E. Chicago," *OGJ*, pp. 96–99, Nov. 1953.
[16] A. W. Sloley, "Consider VFDs for Centrifugal Pumps," *Chem Proc*, vol. 76, no. 4, pp. 44–45, Apr. 2014.

16. Tables and Figures

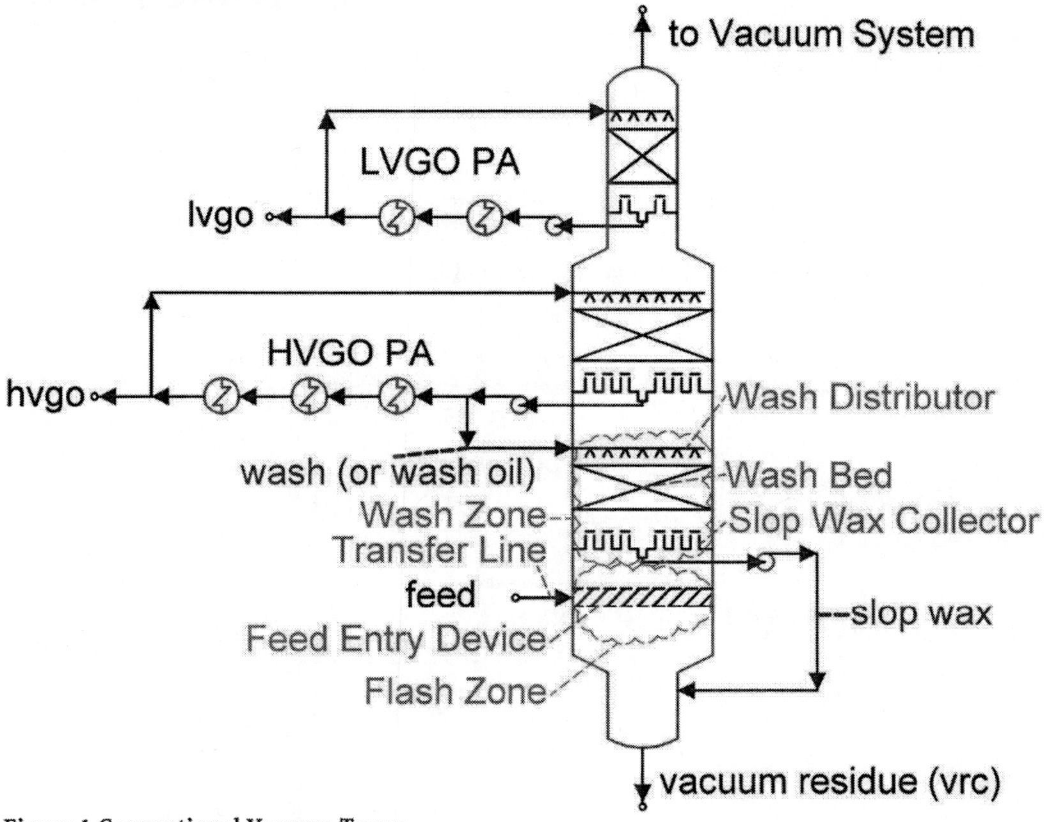

Figure 1 Conventional Vacuum Tower

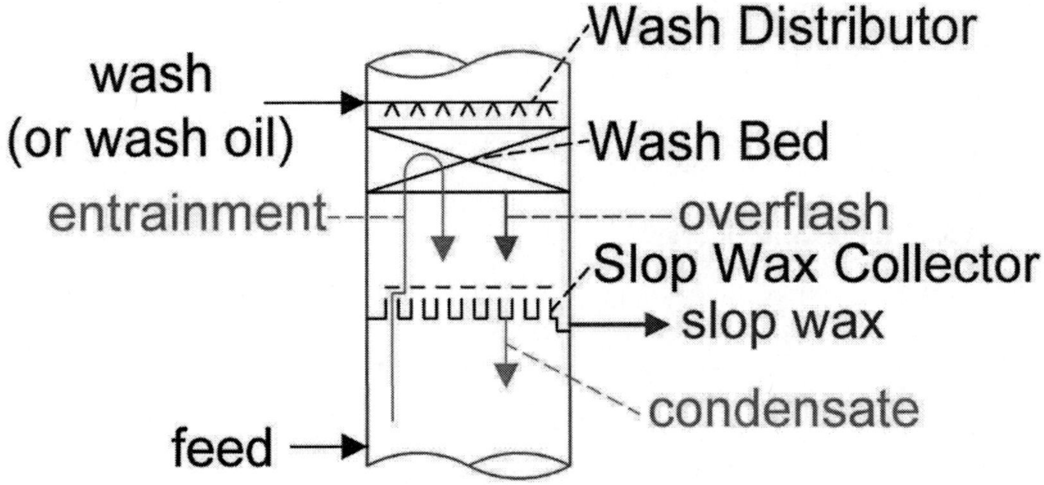

Figure 2 Wash Zone Configuration and Terminology

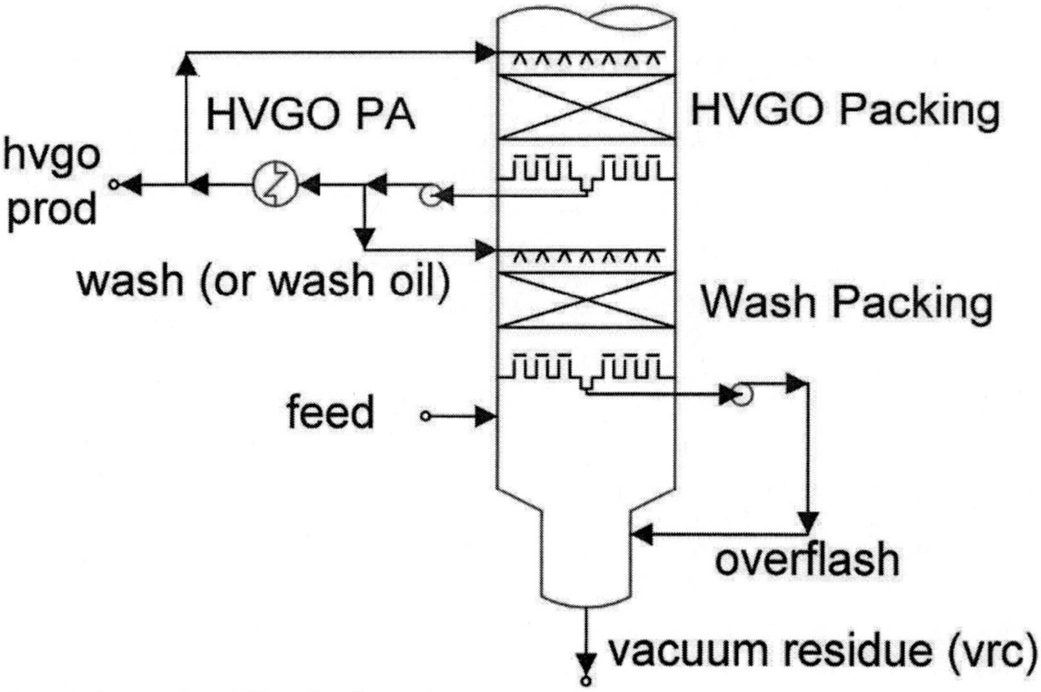

Figure 3 Conventional Wash Configuration

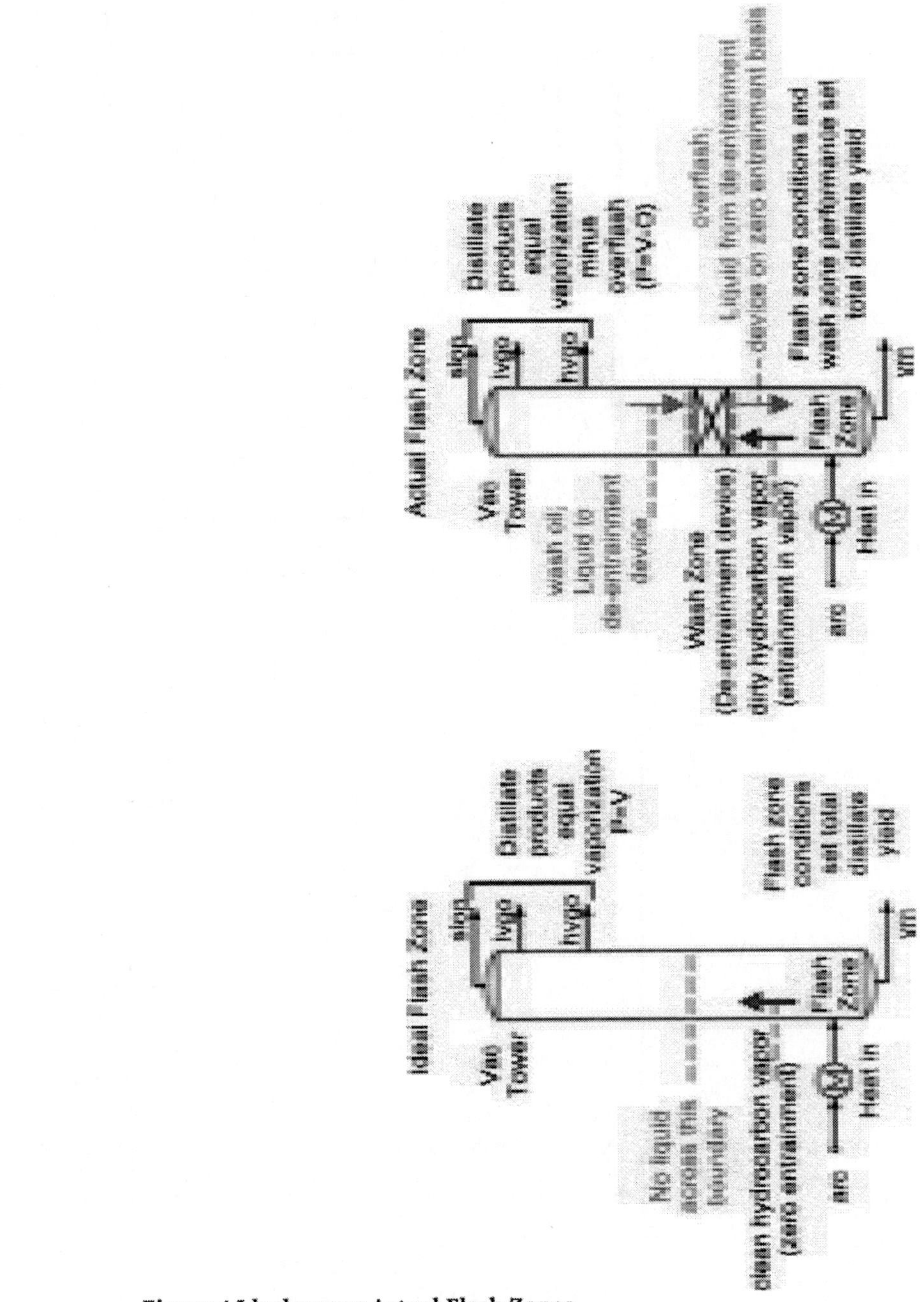

Figure 4 Ideal versus Actual Flash Zones

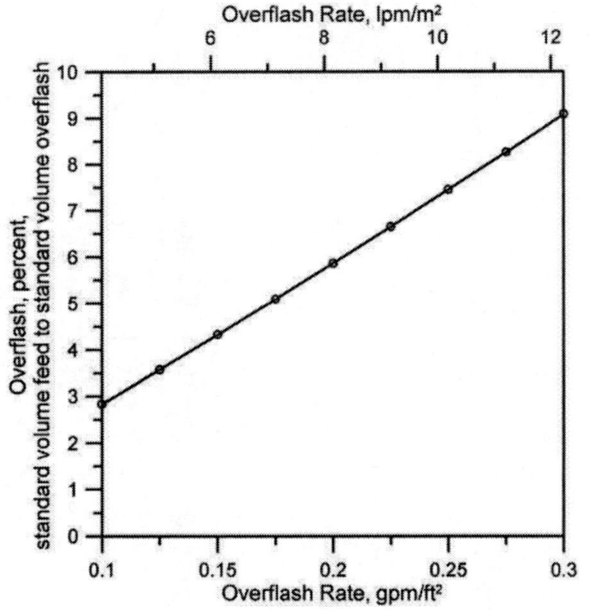

Figure 5 Overflash Percent versus Overflash Rate

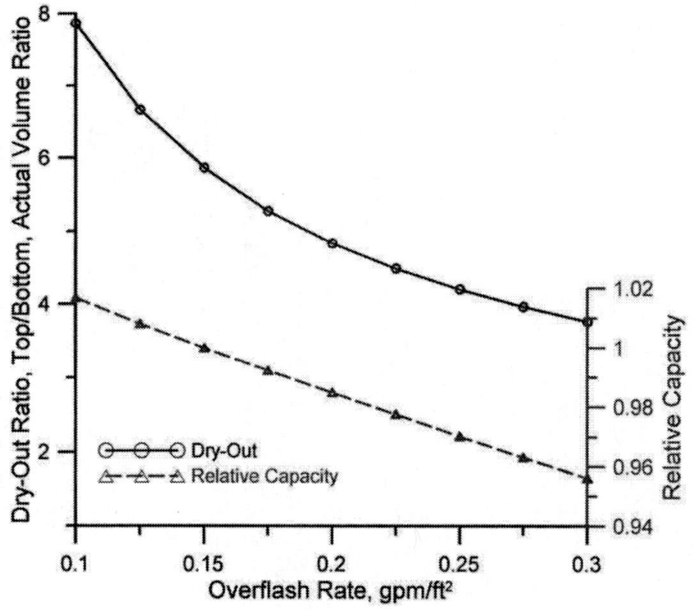

Figure 6 Dry-Out and Capacity versus Overflash Rate

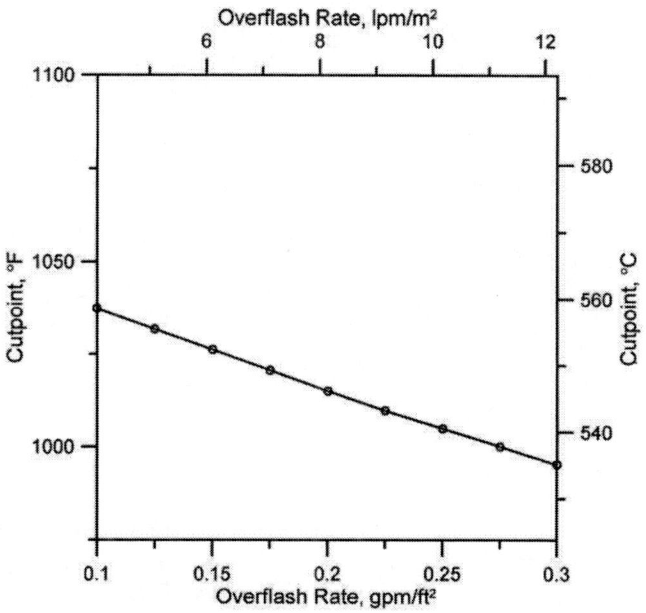

Figure 7 Cutpoint versus Overflash Rate

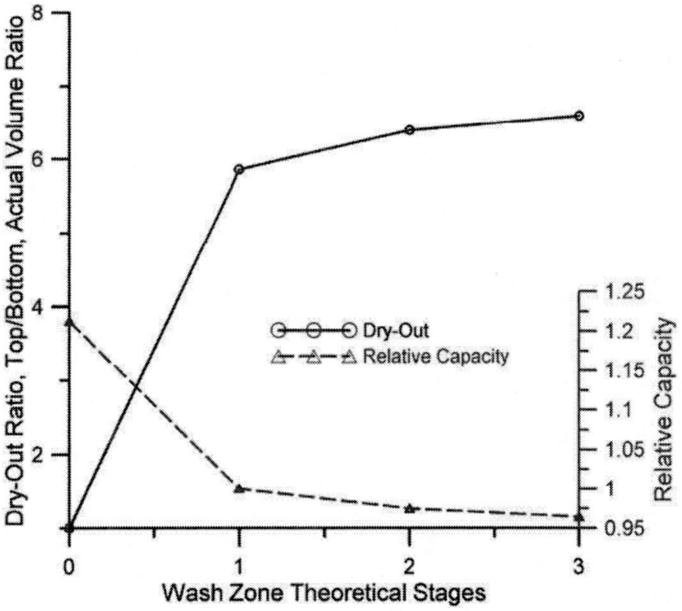

Figure 8 Dry-out and Capacity versus Wash Zone Stages

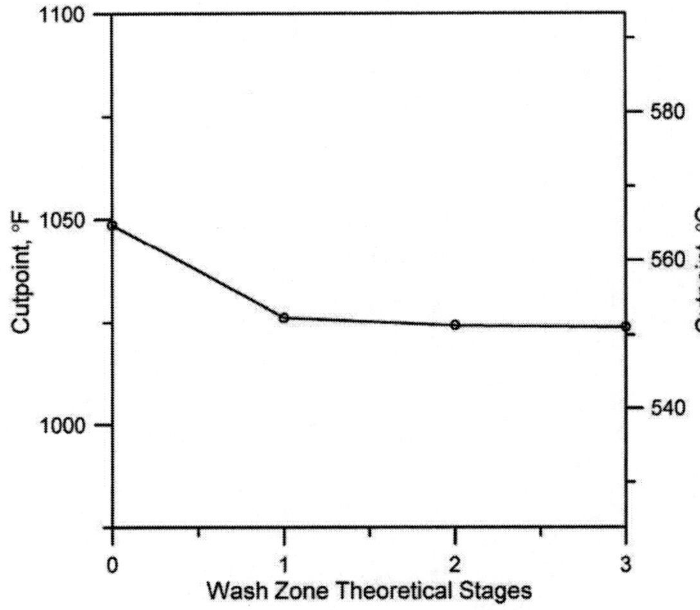

Figure 9 Cutpoint versus Wash Zone Stages

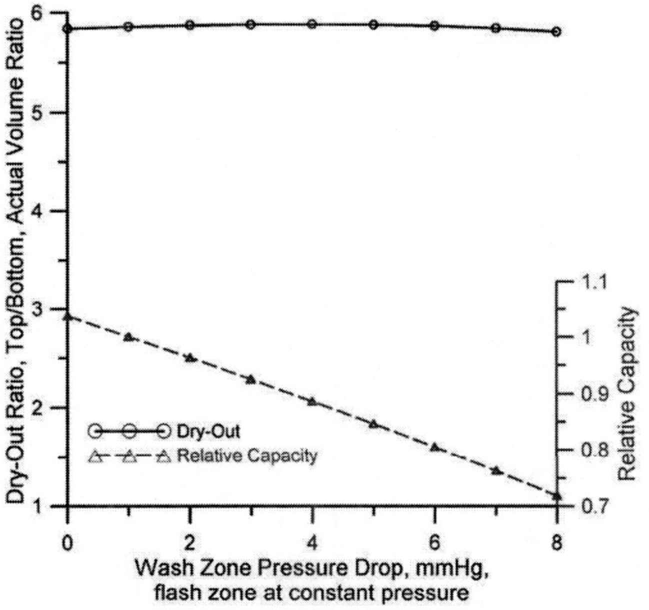

Figure 10 Dry-Out and Capacity versus Wash Zone Pressure Drop

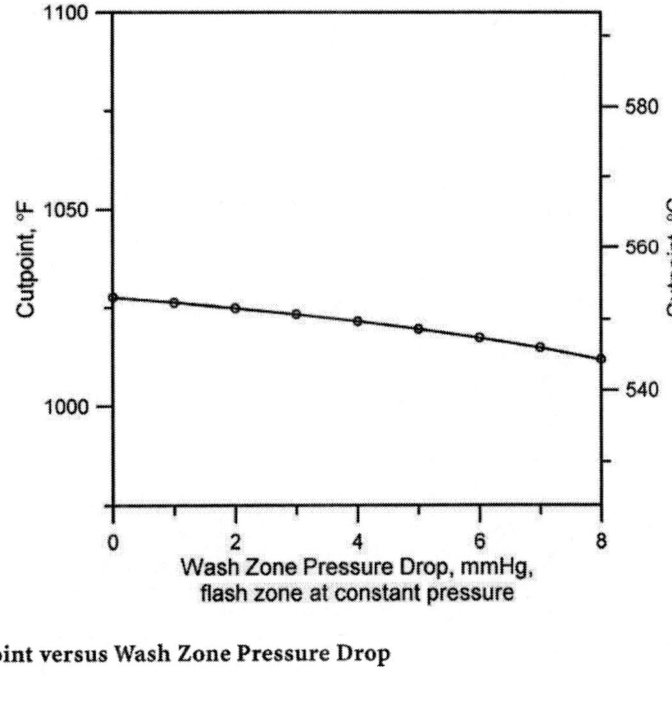

Figure 11 Cutpoint versus Wash Zone Pressure Drop

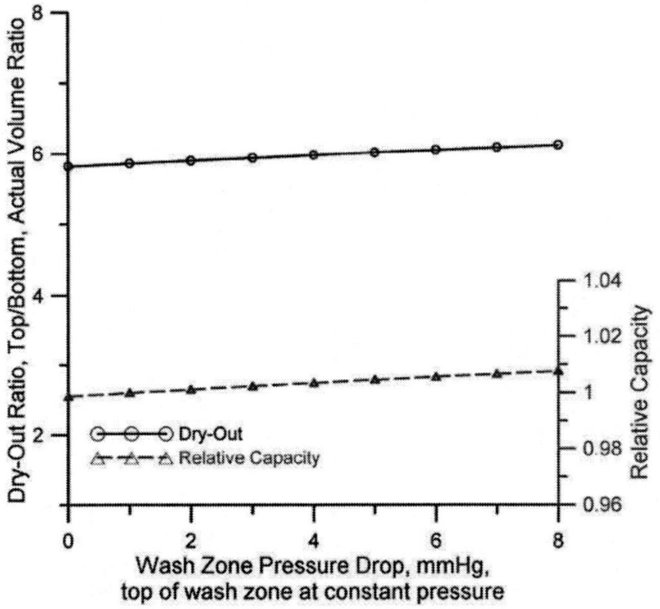

Figure 12 Dry-Out and Capacity Versus Wash Zone Pressure Drop

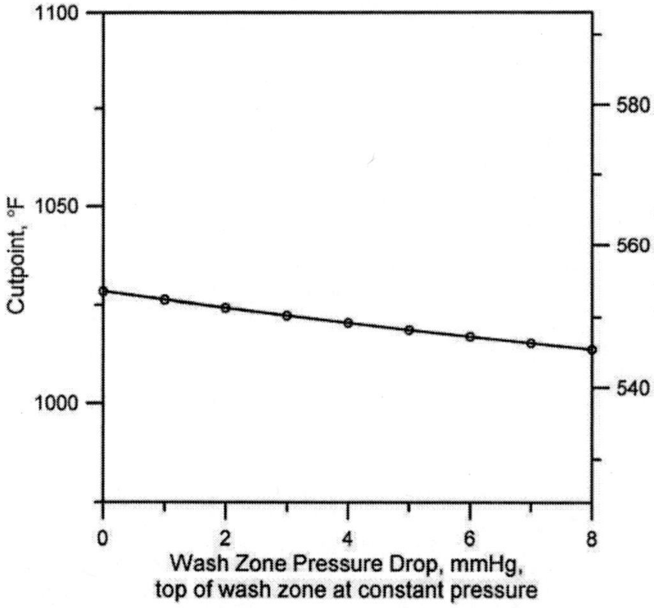

Figure 13 Cutpoint Versus Wash Zone Pressure Drop

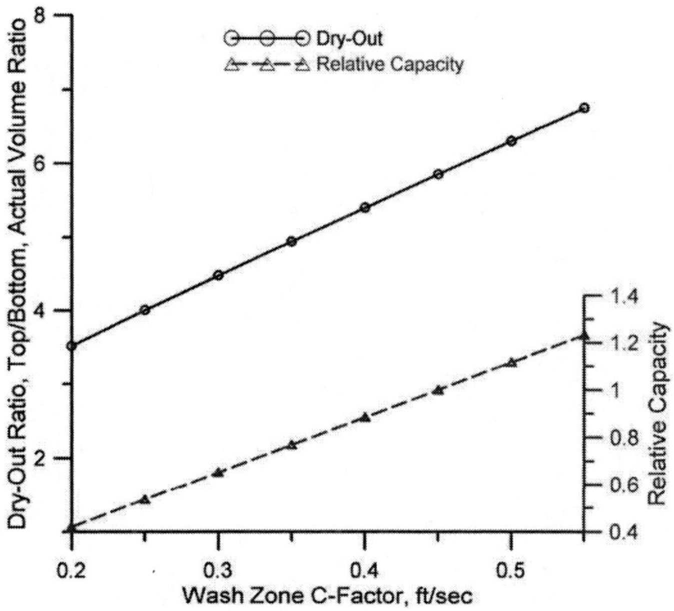

Figure 14 Dry-out Ratio and Capacity Versus C-factor (Constant Feed Rate)

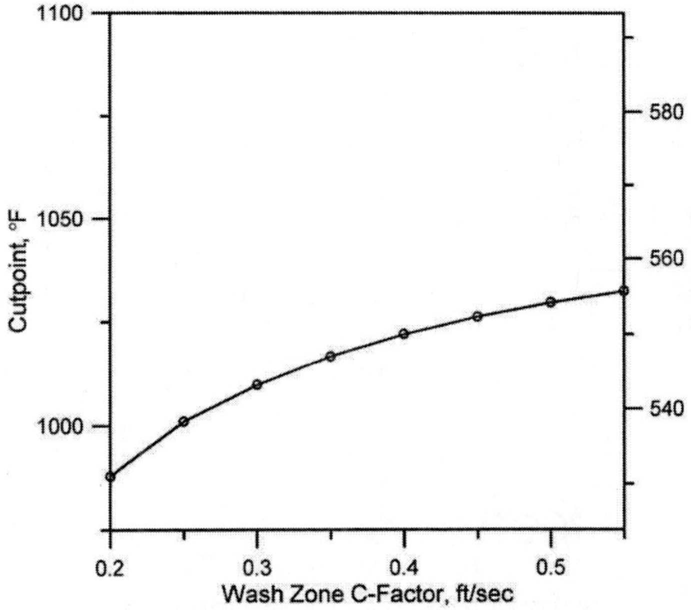

Figure 15 Cutpoint Versus C-factor (Constant Feed Rate)

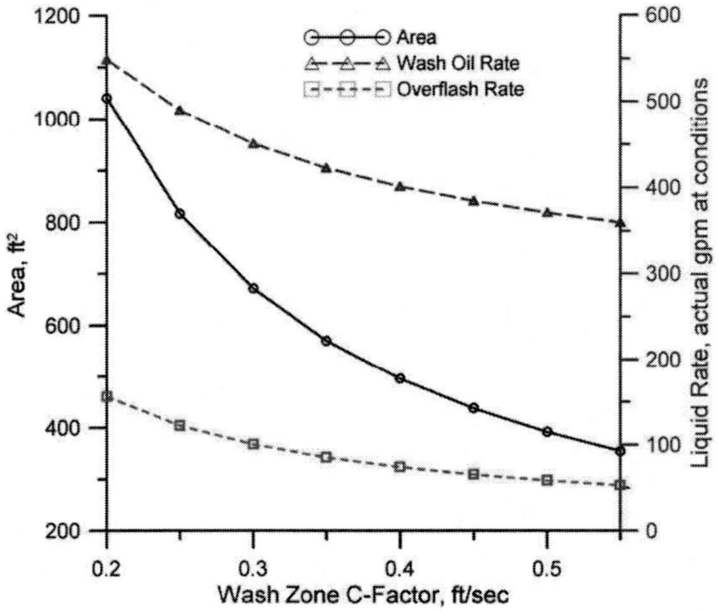

Figure 16 Area and Liquid Rates Versus C-factor

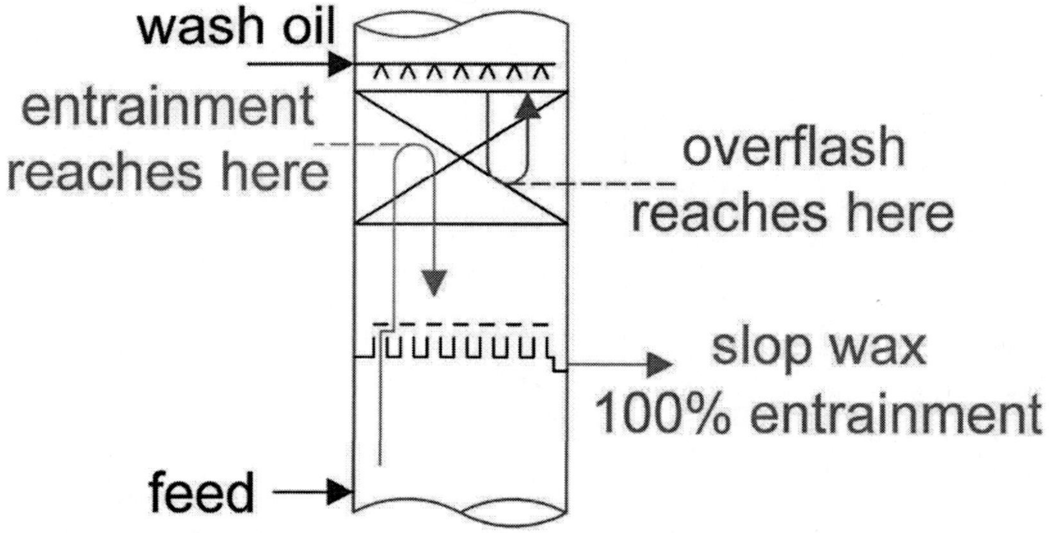

Figure 17 Zero Overflash Operation: Entrainment and Overflash Overlap

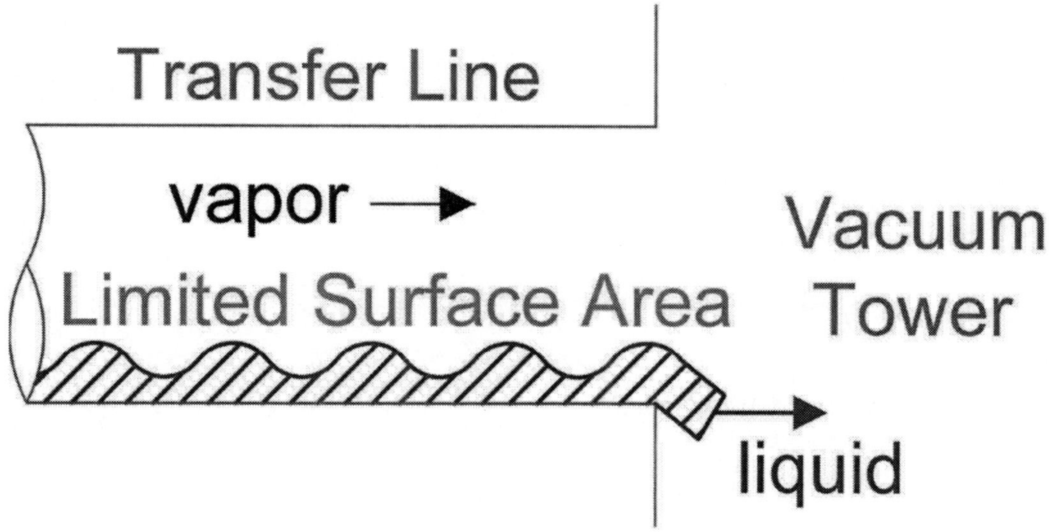

Figure 18 Transfer Line Phase Separation

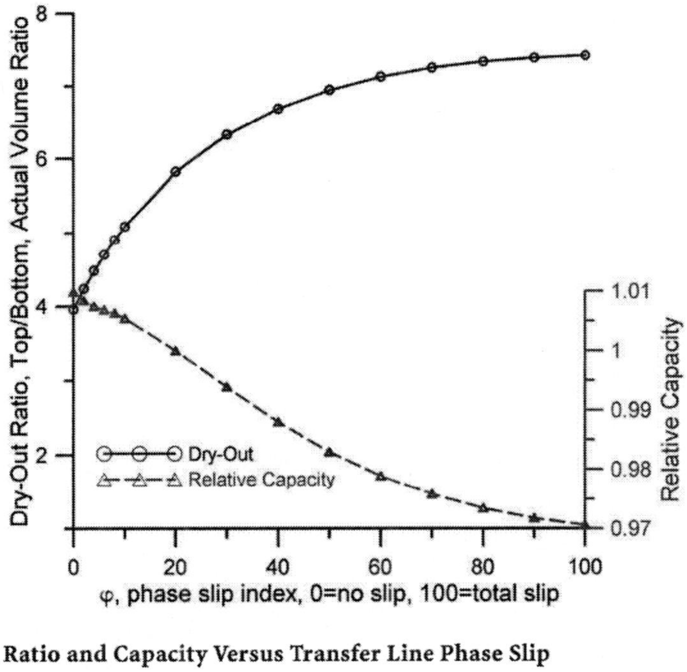

Figure 19 Dry-out Ratio and Capacity Versus Transfer Line Phase Slip

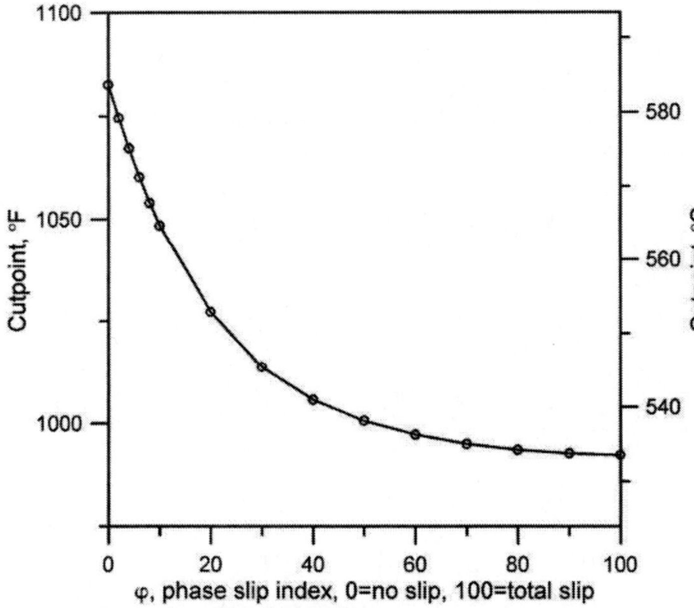

Figure 20 Cutpoint Versus Transfer Line Phase Slip

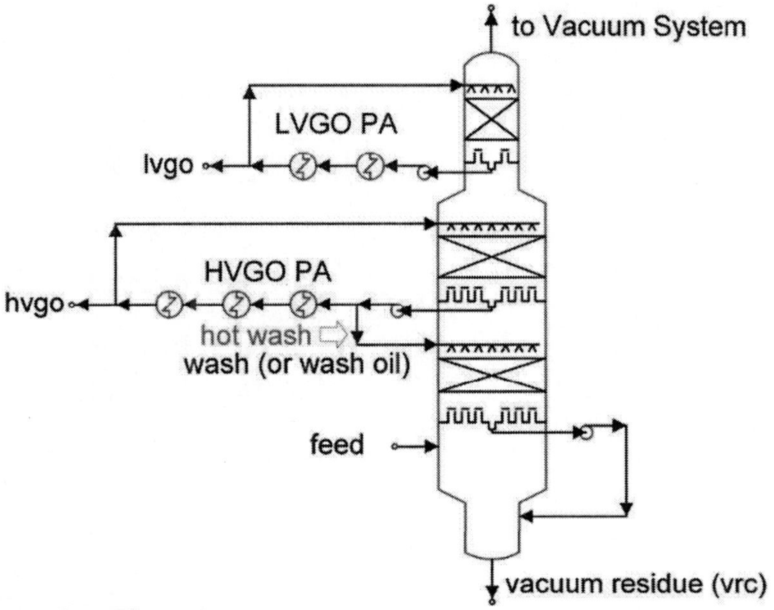

Figure 21 Hot Wash Configuration

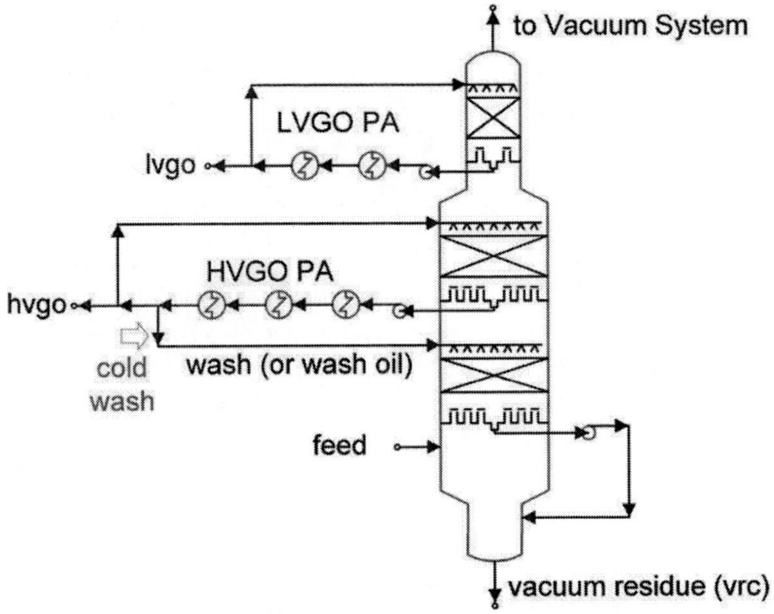

Figure 22 Cold Wash Configuration

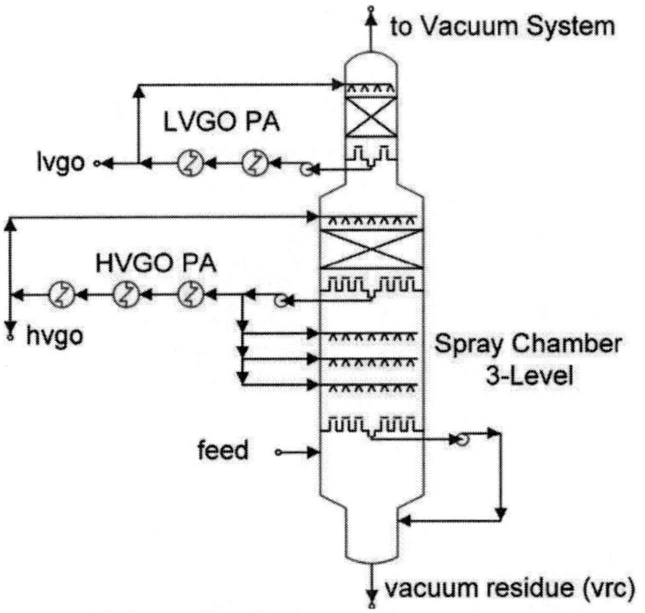

Figure 23 Vacuum Tower With Spray Chamber

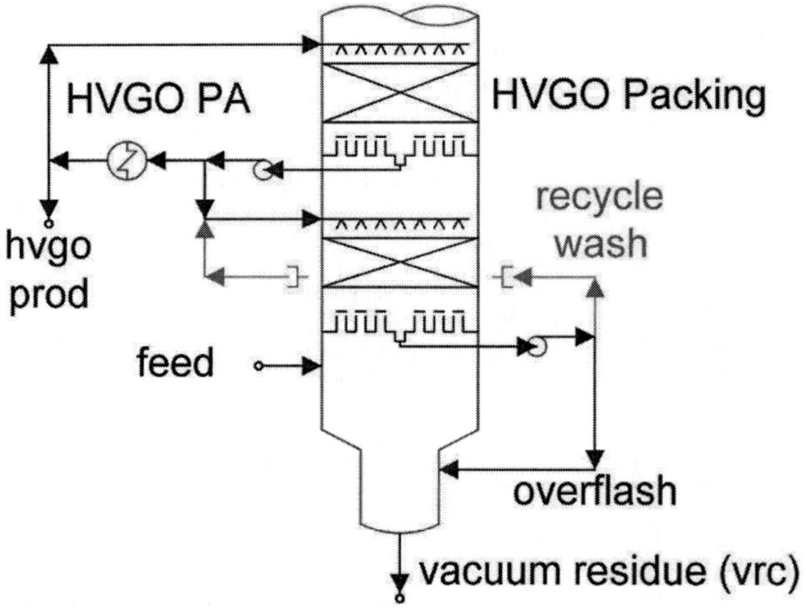

Figure 24 Recycle Wash System: One Stage

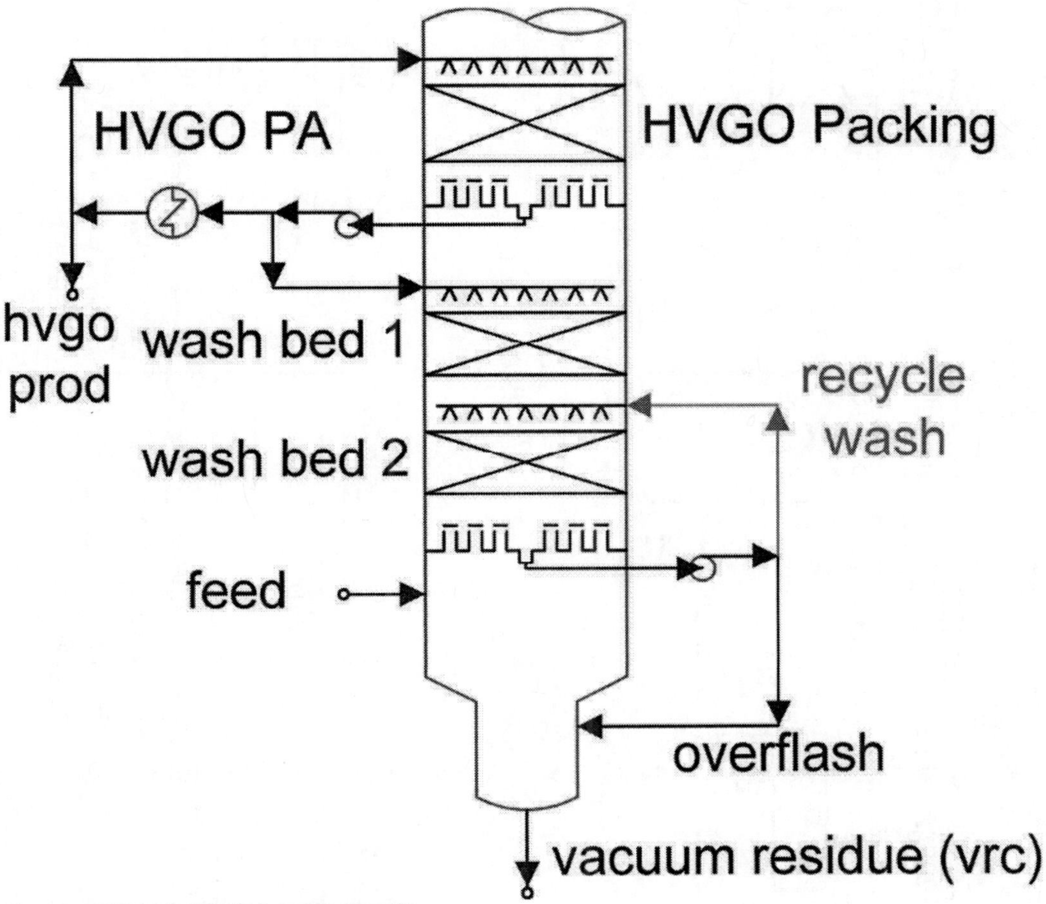

Figure 25 Recycle Wash System: Two Stage

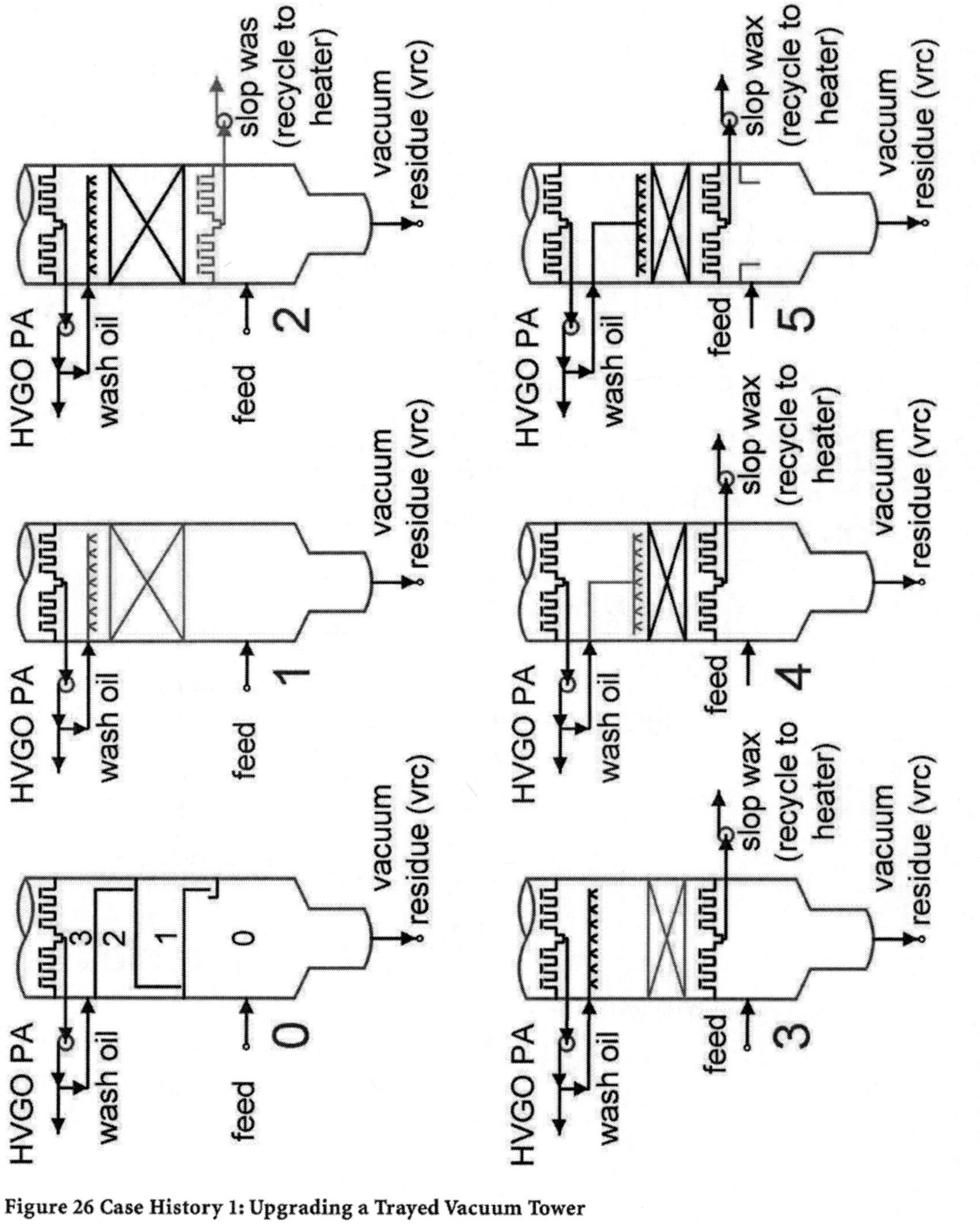

Figure 26 Case History 1: Upgrading a Trayed Vacuum Tower

Optimum Pumparound Balance between Energy & Fractionation Efficiency

Soun Ho Lee – GTC Technology US LLC
2017 AIChE Spring Meeting / Kister Distillation Symposium
San Antonio TX March 27, 2017

+ Engineered to Innovate

Refinery Multi-product Fractionator Pumparound Method

+ Engineered to Innovate

- Refinery multi-product fractionator
  - Multiple product draws
  - Crude distillation unit, FCC(RFCC), hydrocracker, coker
- Pumparound circuits
  - The most common inter-condenser design
  - Minimizing traffic variation
  - Maximizing heat recovery
  - Overhead condenser size reduction
  - Additional trays/packing for heat transfer: higher column T-T elevation

Pumparound Balance
Fractionation vs. Energy Efficiency

GTC Technology – Engineered to Innovate

- Improving fractionation efficiency
- Downgrading energy efficiency

- Improving energy efficiency
- Downgrading fractionation efficiency

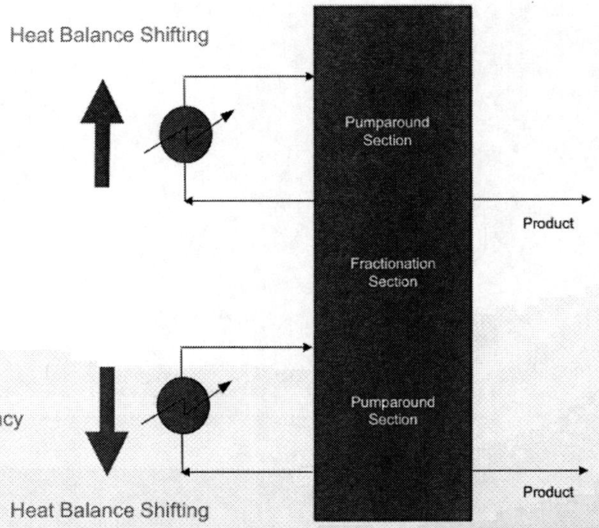

Refinery Multi-product Fractionator
Pumparound / Fractionation Optimization

GTC Technology – Engineered to Innovate

- Non-optimum pumparound balance can be formed if actual process conditions are largely changed.
- Non-optimized pumparound balance can influence performance adversely
- Optimum pumparound balance necessary for target performance

Case Study - Crude Distillation Unit
Pre-revamp Unit Configuration

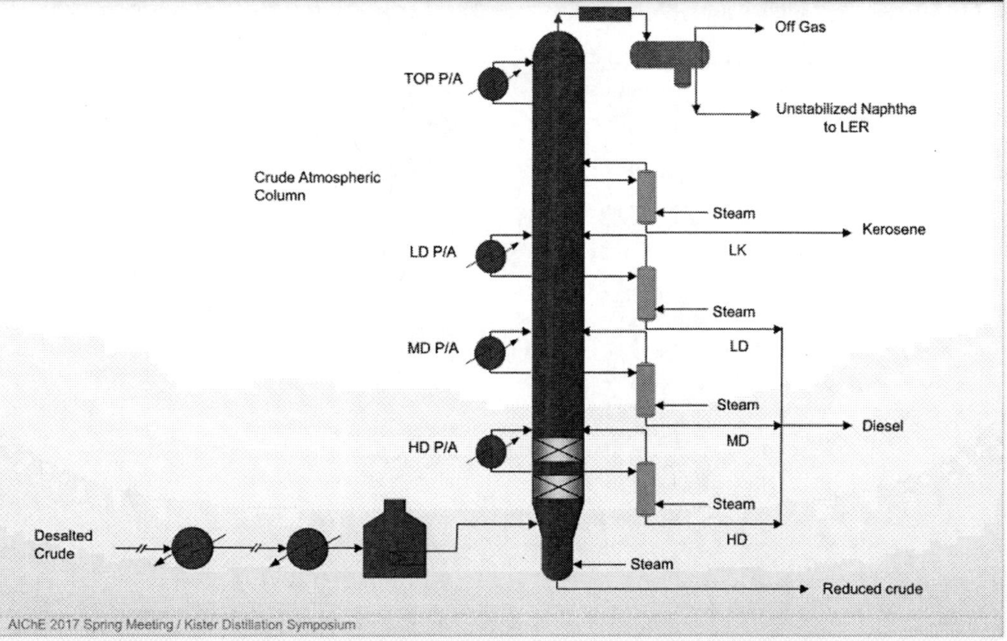

AIChE 2017 Spring Meeting / Kister Distillation Symposium

Case Study - Crude Distillation Unit Background / Pre-Revamp Test Run

- Limitation
  - Kerosene yield limitation while crude slate containing high kerosene boiling range materials was processed
- Test run basis
  - Crude slate: containing high kerosene boiling range materials
  - Charge rate: maximum without fractionation loss
  - ASTM D-7169 for HD and reduced crude stream
- Observation
  - High pressure drop through top P/A and HD P/A section

Case Study - Crude Distillation Unit Feed Composition Comparison

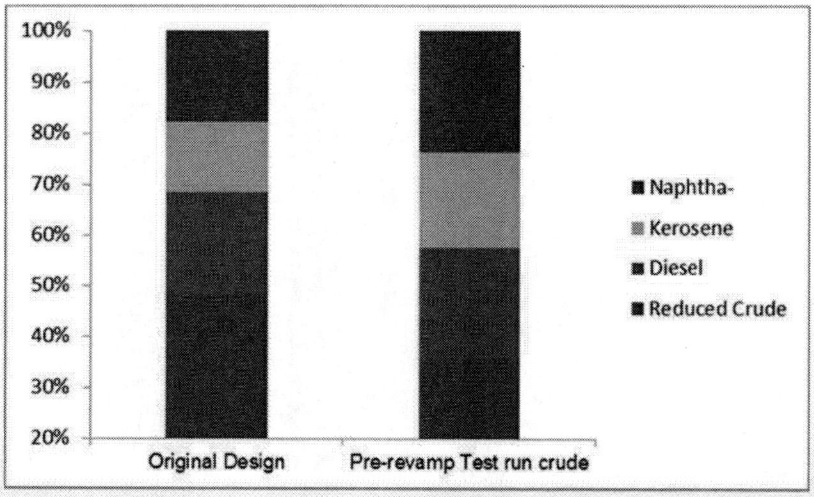

Case Study - Crude Distillation Unit
Pre-Revamp Pumparound Capacity Profile

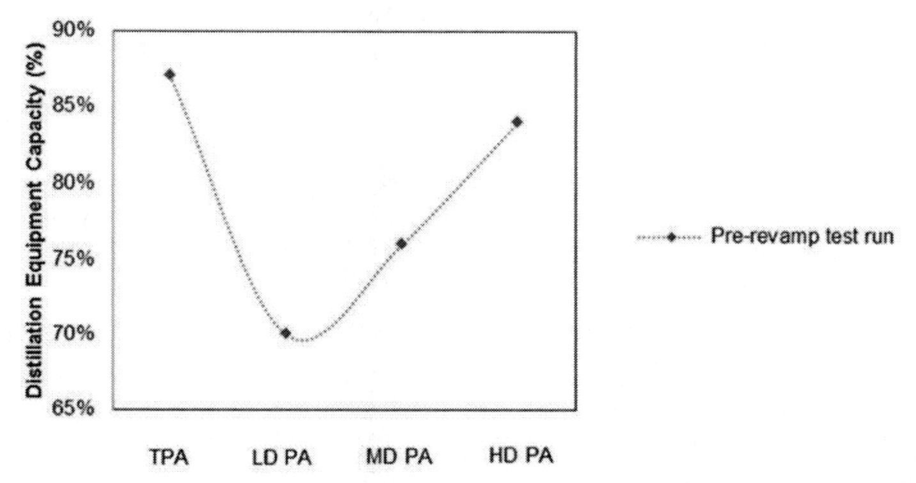

Case Study - Crude Distillation Unit
Revamp Summary

- Re-arranging pumparound circuits for actual feed
  - Mismatch between pumparound locations & feed composition
  - LD P/A was converted to Heavy Kerosene P/A
- Pre-heat train performance was reviewed per new heat balance
  - Heat exchanger modification
- Crude Atmospheric Column
  - Implementing GT-OPTIM^{TM} High Performance Trays

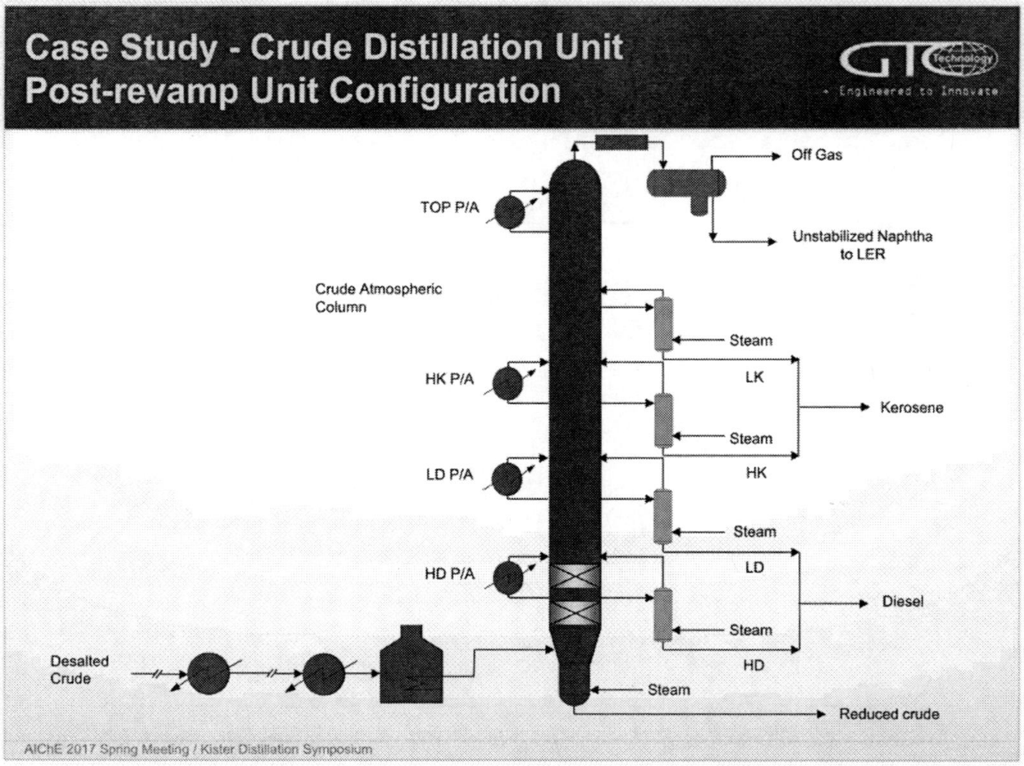
Case Study - Crude Distillation Unit
Post-revamp Unit Configuration
GT Technology
- Engineered to Innovate
Off Gas
TOP P/A
Unstabilized Naphtha
to LER
Crude Atmospheric
Column
Steam
HK P/A
LK
Kerosene
Steam
LD P/A
HK
Steam
HD P/A
LD
Diesel
Steam
Desalted
Crude
HD
Steam
Reduced crude
AIChE 2017 Spring Meeting / Kister Distillation Symposium

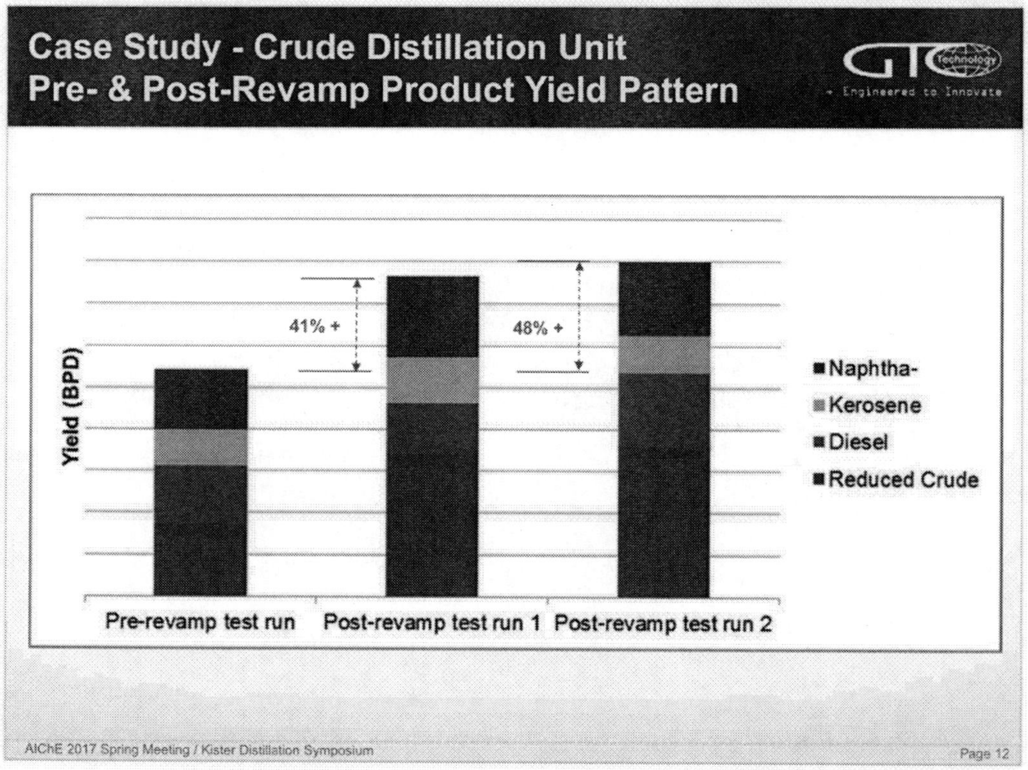
Case Study - Crude Distillation Unit
Pre- & Post-Revamp Product Yield Pattern
GT Technology
- Engineered to Innovate
Yield (BPD)
41% +
48% +
Naphtha-
Kerosene
Diesel
Reduced Crude
Pre-revamp test run
Post-revamp test run 1
Post-revamp test run 2
AIChE 2017 Spring Meeting / Kister Distillation Symposium
Page 12

Case Study - Crude Distillation Unit
Pre- & Post-Revamp Distillate Yield Pattern

GTC Technology – Engineered to Innovate

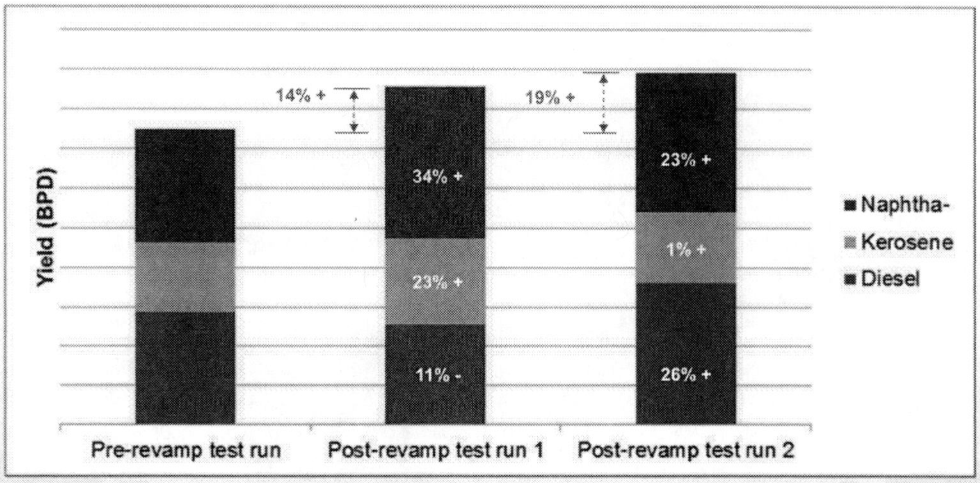

Case Study - Crude Distillation Unit
Pumparound Capacity Profile

GTC Technology – Engineered to Innovate

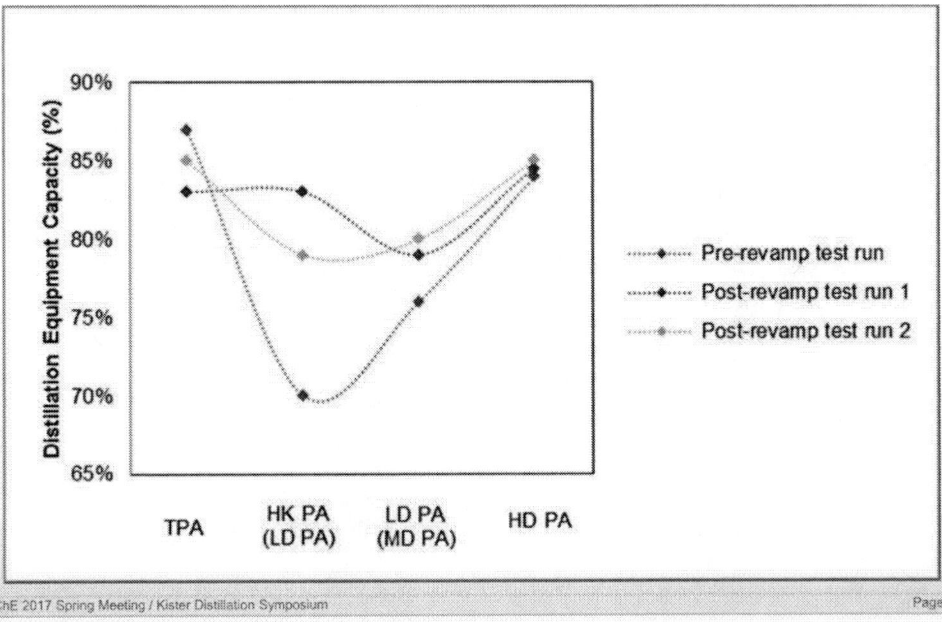

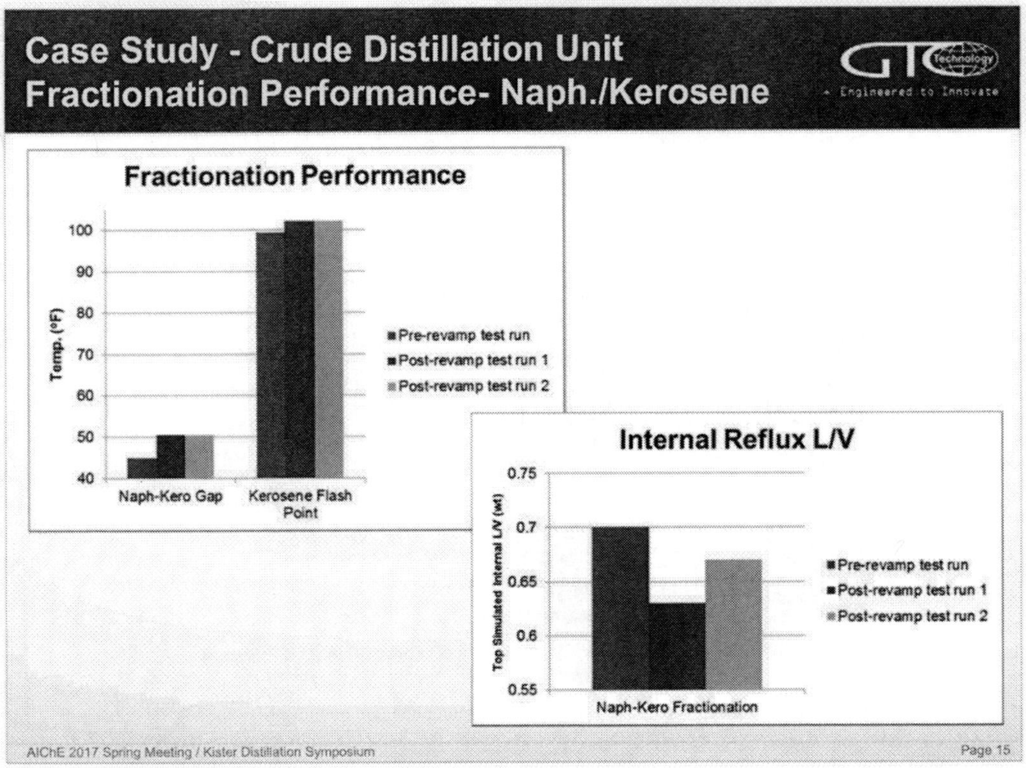
Case Study - Crude Distillation Unit
Fractionation Performance- Naph./Kerosene
GTC Technology
Engineered to Innovate
Fractionation Performance
Temp. (°F)
100
90
80
70
60
50
40
Naph-Kero Gap
Kerosene Flash Point
Pre-revamp test run
Post-revamp test run 1
Post-revamp test run 2
Internal Reflux L/V
Top Simulated Internal L/V (wt)
0.75
0.7
0.65
0.6
0.55
Naph-Kero Fractionation
Pre-revamp test run
Post-revamp test run 1
Post-revamp test run 2
AIChE 2017 Spring Meeting / Kister Distillation Symposium
Page 15

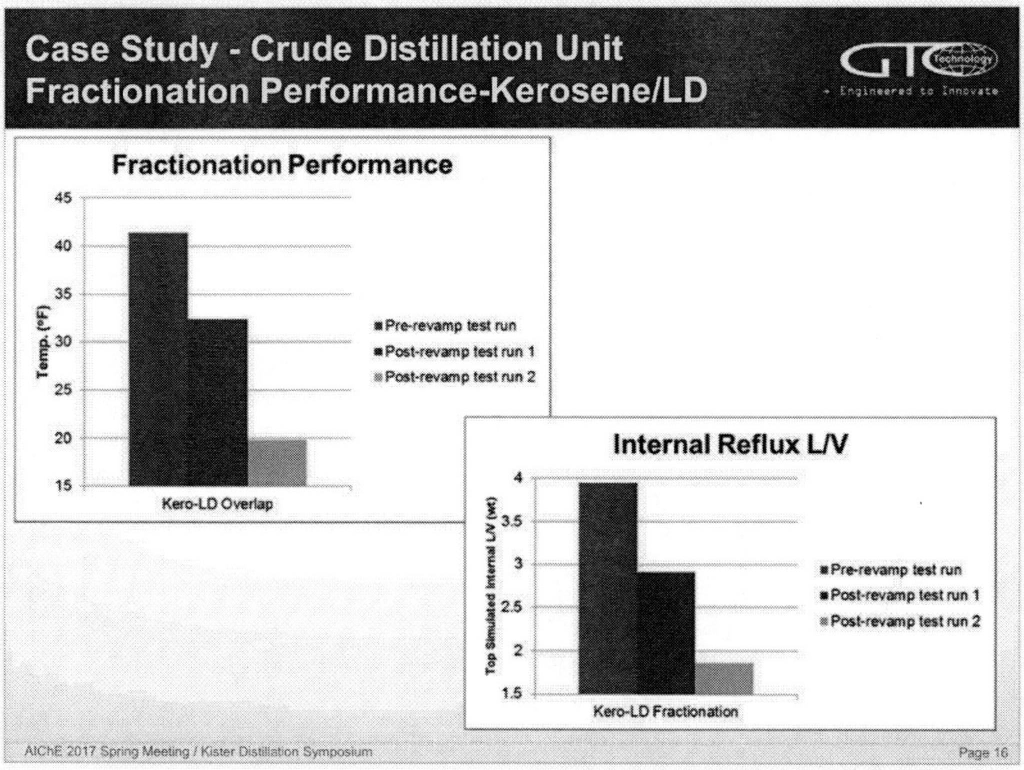
Case Study - Crude Distillation Unit
Fractionation Performance-Kerosene/LD
GTC Technology
Engineered to Innovate
Fractionation Performance
Temp. (°F)
45
40
35
30
25
20
15
Kero-LD Overlap
Pre-revamp test run
Post-revamp test run 1
Post-revamp test run 2
Internal Reflux L/V
Top Simulated Internal L/V (wt)
4
3.5
3
2.5
2
1.5
Kero-LD Fractionation
Pre-revamp test run
Post-revamp test run 1
Post-revamp test run 2
AIChE 2017 Spring Meeting / Kister Distillation Symposium
Page 16

Case Study - Crude Distillation Unit Fractionation Performance- Diesel / RC

GTC Technology - Engineered to Innovate

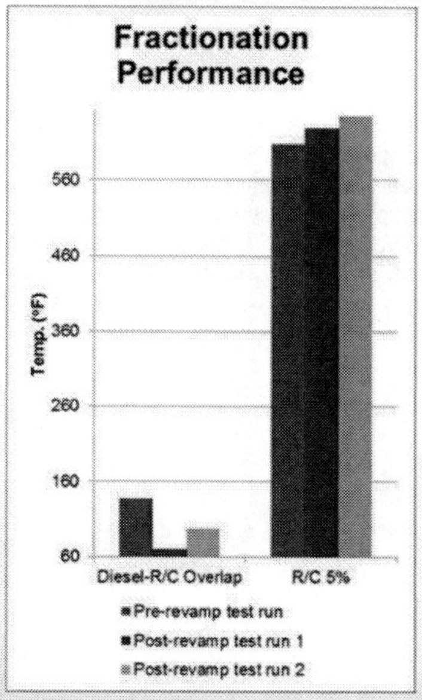

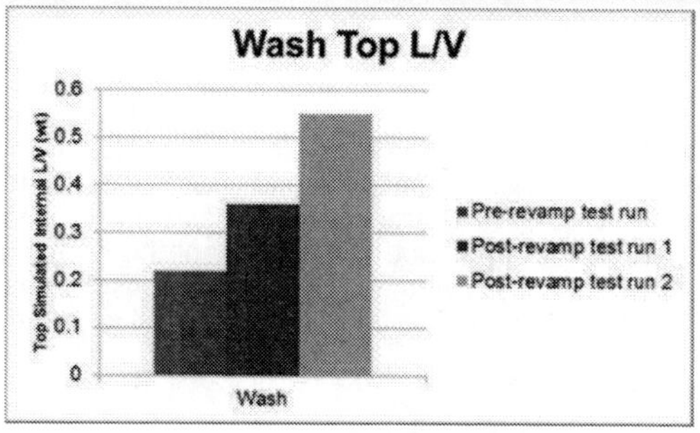

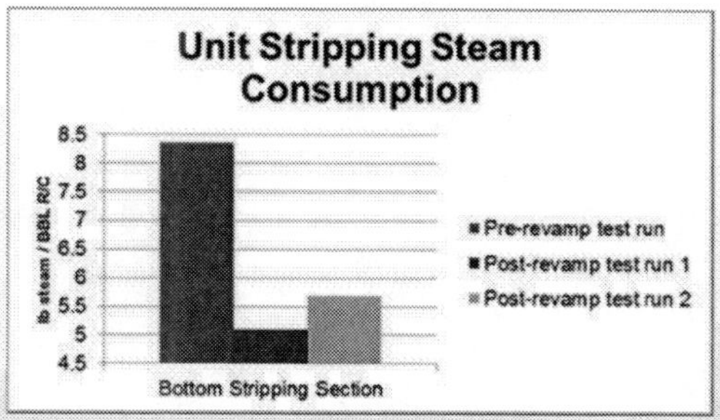

Thank you

GTC Technology - Engineered to Innovate

Thank you

479525 Shell Consep Trays Improve Diesel Recovery and Naphtha Quality in a DHT Stripper

Abstract:

In 2013, Marathon was unable to make naphtha endpoint specifications due to being at 134% of FRI flood in the rectifying section of the low pressure stripper of a diesel Hydrotreater. Since the tower was flooded and the naphtha had such a high endpoint it could not be sent to the reformer, but it was instead recycled back to the crude unit as slop for cleanup. Shell ConSep trays were installed the rectifying section of the tower during the October 2013 Turnaround. Upon startup, naphtha endpoint met specification, which allowed the naphtha recycle to slop to be eliminated. Further optimization of the reflux up to flooding constraints on the tower allowed the naphtha endpoint specification to be decreased, thus increasing the diesel yield. The Shell ConSep trays allowed optimization of the unit with an increased profit of $6.5MM/yr.

481669 Rigorous Troubleshooting Identified Crude and Vacuum Distillation Column Constraints

Abstract:

Valero Ardmore refinery operates a 90 MBD Crude and Vacuum distillation unit, charging primarily mid-continent crude through the Cushing terminal. The heavy products from this unit are routed directly a gas oil hydrotreater (CFHT) or a blowstill tower for asphalt sales. Crude charge capacity became severely constrained 2-years into the operating cycle between turnarounds. Rigorous troubleshooting, including extensive isotope scanning, identified three separate and unrelated areas of limitation in the crude column and one area in the vacuum tower.

The Ardmore refinery crude tower is a typical atmospheric operation (630degF, 40 psig flash zone) with 42 trays, a packed gas oil wash section, and a shrouded stripping section. Overhead light naphtha is routed to a gas plant and atmospheric tower bottoms are routed partially to at CFHT, with the balance to the vacuum tower. Five side draw products from the crude tower are further processed within the refinery. Severe instability and tray flooding was noted throughout the tower, with specific constraints initiating in the top reflux trays, kerosene pumparound trays, and AGO wash bed. Because the limitations were extended through a very large area of the column, dissecting the data and identifying the true limits was a complicated task. The root cause analyses ultimately pointed to contaminants within the crude feed (phosphorous), inherent design issues with the tower internals, as well as operating practices.

The vacuum tower is operated dry, with a moderate flash zone severity (C-factor <0.4fps). The wash bed reliability is a typical run-limiting constraint to the unit, with wash oil controlled to maintain a slop wax draw rate (metered and pumped to CFHT) indicative of adequate overflash. Loss of slop wax draw occurred at a similar time to the atmospheric constraints. The troubleshooting identified an obstruction on the tray that created residence time and ultimately coke.

In this paper, the authors will describe the troubleshooting methodology, discuss an unconventional hot tap bypass that allowed the unit to remain in operation until a planned turnaround and outline design changes that were implemented during the outage. The authors will also present turnaround findings that prove the conclusions drawn from operational and troubleshooting data analysis were correct.

Correct Problem Diagnosis Debottlenecks a Gas Plant with No Hardware Changes

By

Henry Z. Kister, Fluor, Aliso Viejo, CA, USA

Presented at the Distillation Topical Conference, AIChE Spring Meeting, San Antonio, Texas, March, 2017

Summary

This paper describes a case study of a gas plant (designed by others) that was bottlenecked by the reboil system of a deethanizer stripper. The reboiler system that consisted of two parallel once-through thermosiphon circuits experienced two different steady states. When operated at low rates, an undesirable steady state was established with maldistribution between the two circuits, in which one was circulating and the other remained liquid-full. This steady state persisted when the tower feed rates were raised until the circulating reboiler reached its heat transfer limitation, still with the other circuit liquid-full.

Pressure drop calculations, close design reviews, examination of operating charts, and gamma scans, led to the correct diagnosis that enabled debottlenecking the reboiler circuits and the entire gas plant with no hardware modifications. The desired steady state was established by an operation that temporarily generated higher liquid circulation and pressure drop through the reboiler circuits.

This case teaches a number of very important key lessons:

- With a parallel reboiler system, maldistribution between parallel circuits does occur and can cause severe bottlenecks. To the best of our knowledge, this is the first time that such maldistribution is reported in the literature.
- Pressure balances need to be considered at the design not only for the maximum design rates, but also at the minimum expected operation rates.
- It is important to provide flexibility in once-through reboiler systems that will allow operating as a circulating thermosiphon systems, at least at low rates.
- Finally, good troubleshooting can provide low cost (in this case, no cost) debottleneck of an entire gas plant.

Background

This newly-built gas plant was bottlenecked by the deethanizer stripper. The tower bottleneck appeared in the reboiler circuit. In early April, 2014, Fluor, who had no involvement in the tower or reboiler design, but has extensive experience in distillation design and troubleshooting, was requested to assist with investigating the root cause of the bottleneck and in finding ways of overcoming it. A task force of operating company process and operation engineers, and Fluor's distillation specialists, was formed. The joint team identified the root cause. Based on the root cause, a solution was successfully implemented. After this the deethanizer stripper and the gas plant were in full operation achieving the design rates. This paper describes our troubleshooting investigation and the lessons learnt.

Process Description

The deethanizer stripper is a 15'-6" tower equipped with 4-pass conventional valve trays with moving round valves. The tower base contains a preferential baffle that divides the bottom sump into a separate reboiler draw compartment and a separate bottoms draw compartment. Liquid leaving the bottom valve tray is diverted to the reboiler draw compartment (Figure 1). From there it flows into a battery of four horizontal thermosiphon reboilers. The bottom tray liquid is diverted to the reboiler draw compartment by notched bottom seal pans. As shown in Figure 2, the 8" deep notches in the seal pans are located right above the reboiler draw compartment. There are no notches above the bottom draw compartment. The purpose is to feed all the bottom tray liquid to the reboilers.

The vapor-liquid mixture returning from the thermosiphon reboilers enters above the bottoms draw compartment via two opposite 24" nozzles. Each nozzle is equipped with a pipe that extends to near the centreline of the tower and has a 4'-6" cutout in the bottom quadrant that is directed towards the bottom draw sump (Figures 1 and 2). The purpose is to have all the reboiler return liquid enter the bottom draw compartment, and none enter the reboiler draw compartment. The vapor returning from the reboilers ascends to the bottom tray, while the liquid returning from the reboilers descends to the bottoms draw compartment. Any excess bottom tray liquid not boiled by the reboilers overflows the baffle into the bottoms-draw compartment.

This design generates a once-through[1] thermosiphon reboiler system. Such a system intends to capture all of the liquid leaving the bottom tray and feed it directly to the reboiler. The reboiler return liquid is entirely drawn as bottoms product. None of the reboiler return liquid is recycled back to the reboiler, hence the name once-through.

Once-through systems can achieve one full theoretical stage of separation, if the trapout tray and draw pan do not leak[1]. With the cold bottom tray liquid being fed to the reboiler, it can also maximize the reboiler temperature difference and capacity. Reboil vapor is limited to about 40% of the bottoms product rate, due to the normal limitation of 30% maximum vaporization (by weight) in thermosiphon reboilers[1]. This makes once-through thermosyphon systems appropriate only for low-boilup systems such as strippers. Many deethanizer-strippers use once-through thermosiphon reboiler systems.

Figure 1. Deethanizer Stripper Reboiler Circuits

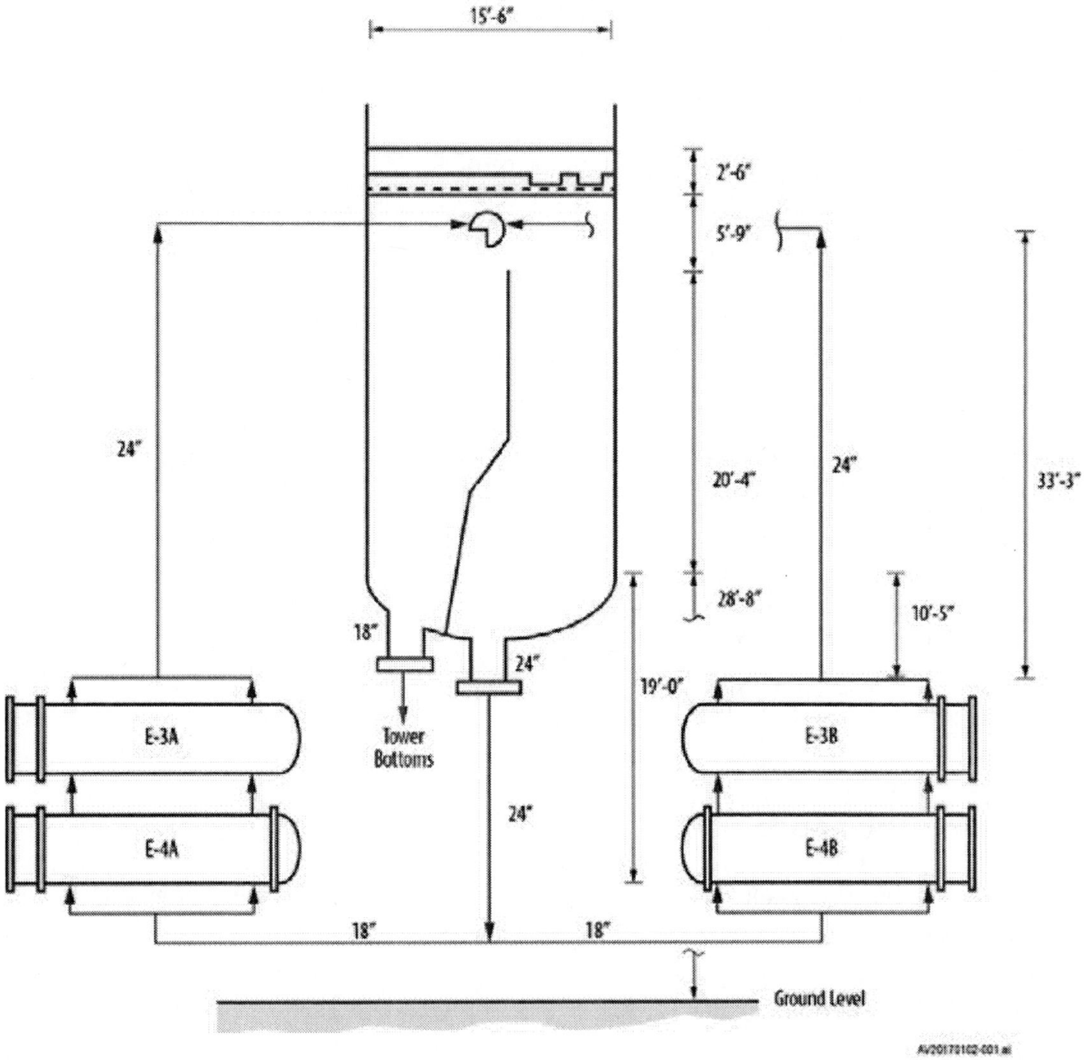

This deethanizer stripper has four horizontal reboilers, arranged in two parallel circuits, "A" and "B" (Figure 1). In each circuit, the lower reboiler (E-4A or E-4B) is heated by hot naphtha liquid, while the upper reboiler (E-3A or E-3B) is heated by high-pressure steam. In each circuit, the naphtha and steam reboilers are designed to provide duties of 32 MM Btu/h and 40 MM Btu/h, respectively. On the boiling side, liquid to the reboilers is provided from the reboiler draw sump via a 24" pipe which splits into two 18" pipes, each going to one circuit. The reboiler effluent from each circuit returns to the tower via a 24" line.

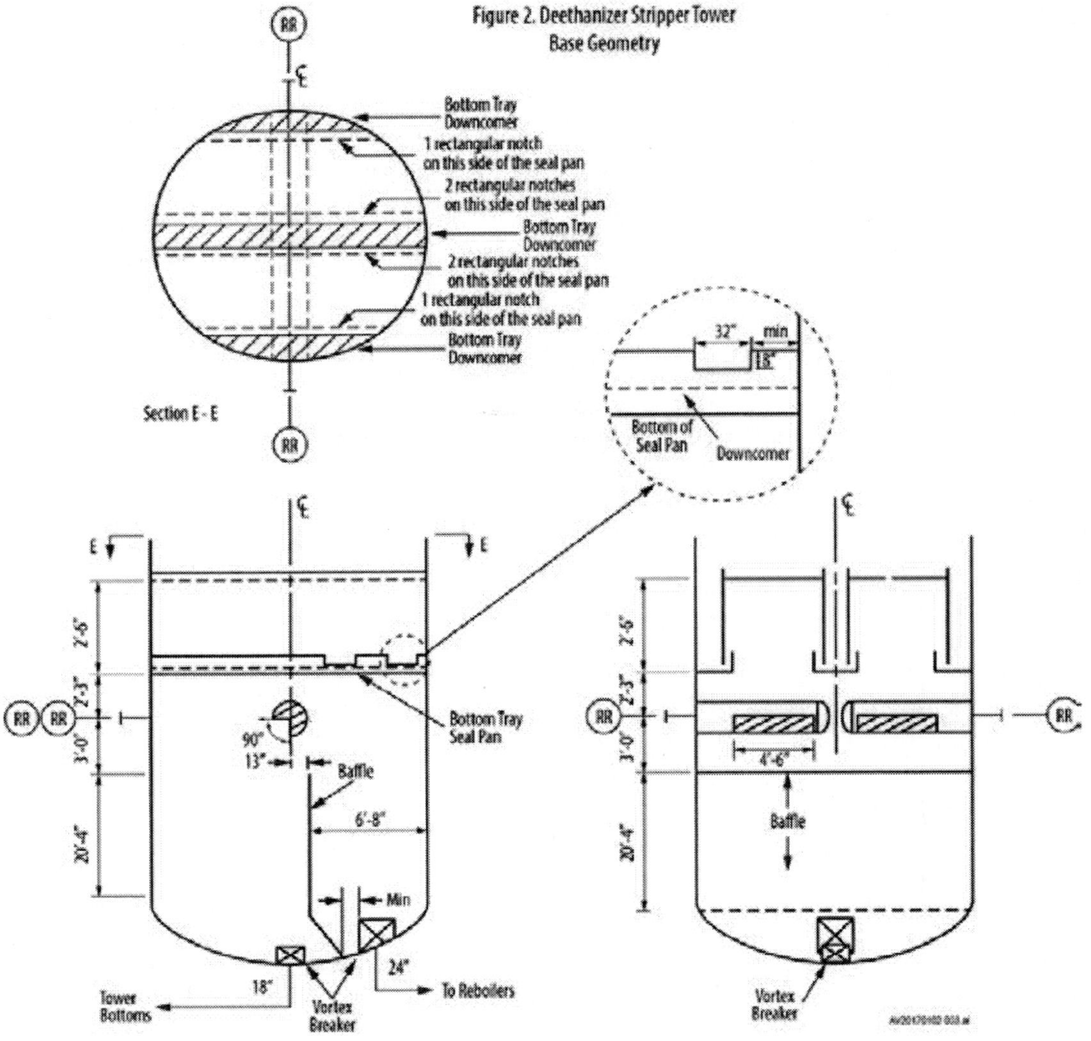

Figure 2. Deethanizer Stripper Tower Base Geometry

Initial Performance

During the start-up, the plant was at low rates and to improve tower stability it was decided to take the steam out of reboiler E-3A. This was successful at creating a steady steam flow through the E-3B reboiler circuit and provided tower stability but the plant has been unable to re-establish flow through the E-3A/E-4A set of reboilers ever since. For the next six months or so, the E-3B/E-4B circuit did all the reboiling, while the deethanizer bottoms liquid circulation has been difficult to establish through the E-4A/E-3A circuit.

Figure 3 shows typical operating data from 6 months after start up and compares them to the design values. The zero steam flow rate through E-3A and the absence of significant temperature difference across the heating side of E-4A indicate no boiling and no significant circulation through the A circuit. To maximize duty, steam to E-3B was raised to the maximum, well above the design steam rate for this exchanger, but still not high enough to make up the lost duty of E-3A. This restricted the feed flow rate through the tower and bottlenecked the entire gas plant. There was a need to get the stripper bottoms liquid to start flowing through the E4A/E-3A circuit.

Figure 3. Deethanizer Stripper
Operation vs. Design
Operation____ (Design____)

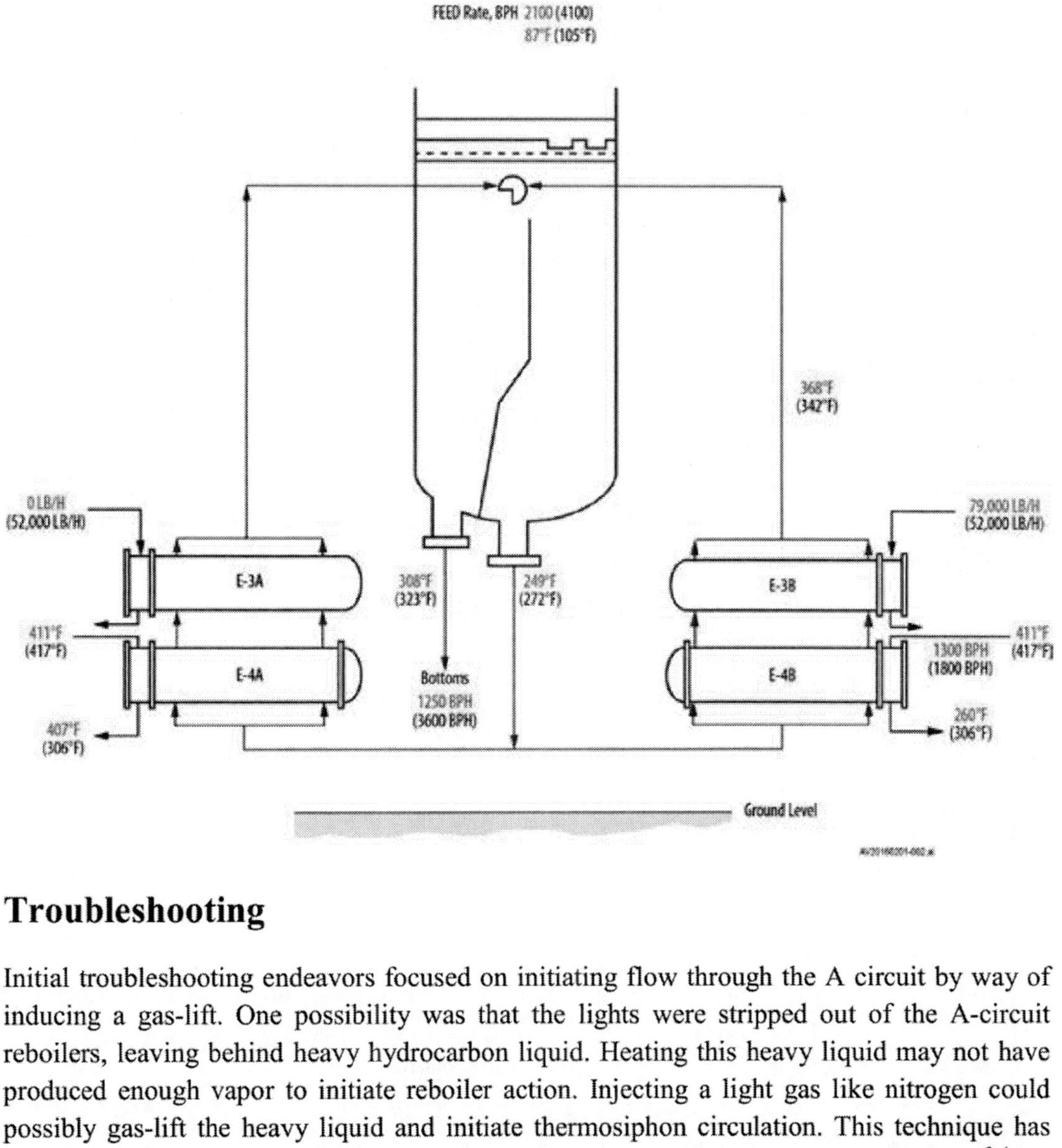

Troubleshooting

Initial troubleshooting endeavors focused on initiating flow through the A circuit by way of inducing a gas-lift. One possibility was that the lights were stripped out of the A-circuit reboilers, leaving behind heavy hydrocarbon liquid. Heating this heavy liquid may not have produced enough vapor to initiate reboiler action. Injecting a light gas like nitrogen could possibly gas-lift the heavy liquid and initiate thermosiphon circulation. This technique has been successful in the past for initiating circulation in liquid-full thermosiphon reboilers[2-4].

A source of nitrogen was connected to a bleeder valve immediately downstream of E-3A. Nitrogen was injected for 4-6 hours in an attempt to provide lift and jump-start the thermosiphon effect. This action was unsuccessful in initiating circulation through the A circuit.

This came as a surprise, as in other reboilers[2-4] a similar action had been successful. There were questions as to whether enough nitrogen was applied, whether it was applied at the correct point, and whether the liquid in the reboilers was just too heavy and needed draining and replacing by fresh liquid before the gas lift attempt. Addressing these problems would have required some painful modifications or operations with no assurance of success. The validity of the theory that heavy liquids in the A-circuit are the root cause was questioned. With these in mind it was decided to perform more troubleshooting before trying out a fix.

One useful reference point was process data from the very initial operation of the stripper. The operating charts from shortly after initial start up (Figure 4), at a stripper feed rate of about 1700 bph, show both the A and B circuits operating simultaneously and stably. Both E-3A and E-3B reboilers operated stably with a steam flow rate of about 32,000 lb/h to each. For both the E-4A and B reboilers, the temperatures of naphtha entering and leaving were about 400°F and 320°F, respectively. At 22:40 the stripper feed rate plunged, steadying at about 1200 bph forty minutes later. To improve stability, the plant manually reduced the steam flow rate to E-3A to about 8000 lb/h, with the steam rate to E-3B remaining at its previous value. The naphtha temperatures into and out of E-4A became the same, about 360°F, suggesting it stopped heating. At the same time the temperature outlet from E-4B decreased to about 260°F, suggesting that this reboiler was working harder than before. Both of these suggest that the A-circuit stopped circulating. Early on the next day, the feed rate to the stripper rose again, to 1500 bph, but this time the A circuit remained inactive. It was common practice to attempt to restart circulation by opening and closing the steam valve on the stalled set but circulation remained stalled.

Figure 4. Process Data for Initial Operations of Stripper Reboiler Circuits

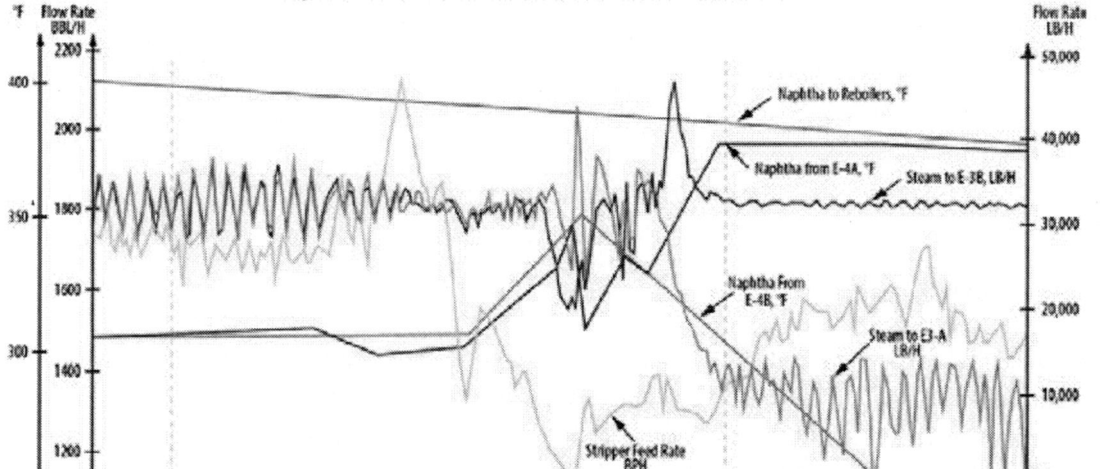

This sequence of events suggests the possibility of the existence of two hydraulic steady states, with an easy switchover from one to the other. This is likely to take place when the reboilers pressure drop on the boiling side is low.

Calculations performed during the design by the engineering contractor that designed the circuits showed a total pressure drop of 1 psi for the exchangers in each circuit at the design conditions. The contractor's calculated pressure drop were 0.4 psi and 1.3 psi for the reboiler liquid inlet line and the reboiler return lines, respectively. The total pressure drop is equivalent to 11 feet of liquid head.

Figure 1 shows the elevations of the reboilers circuits. The diagram shows that not much liquid head is required to overcome the pressure drop in the reboilers circuit. The liquid level in the reboiler draw compartment will therefore be near the bottom of the compartment, even at the design rates. At the low rates, the pressure drop will be lower still, and the liquid level will drop further, possibly dropping out of the reboiler draw compartment into the reboiler liquid inlet pipe. Further, at low boilup rates the bottom tray may weep, sending liquid directly into the bottom draw compartment, and causing the liquid level to be lower still.

At a low liquid head, fluids tend to form their favorite flow patterns. Two steady states can form: one, with operation of the two reboilers simultaneously, as intended. The other has all the boiling taking place in one circuit, and the other circuit filling with liquid to the same height as the liquid in the reboiler liquid inlet pipe. Once this steady state forms, the steam that is still going in (the 8000 lb/h going to E-3A after midnight on the second day in Figure 4) will strip the lights out of the near-stagnant liquid. Before long, heavy liquid will form in this circuit, and this liquid will be difficult to boil. This will stabilize this steady state and it will persist.

The proposed theory states that the liquid level in the reboiler compartment would be very low, possibly absent, and will not approach the overflow baffle. This can be easily tested using gamma scans. The tower base was gamma scanned both on the bottom compartment side and on the reboiler compartments side. In addition, neutron backscatter scans were performed to confirm or deny the gamma scan results.

The gamma and neutron scans showed the following:

- Nothing abnormal showed up on the inner active area scans for the bottom 6 trays. Other trays were not scanned.
- Both the gamma and neutron scans showed that the liquid level in the bottom draw compartment was 10-12 feet above the tangent line. This was the same as the level indicated by the level instrument.
- Both gamma and neutron scans had difficulty finding a liquid level on the reboiler draw compartment. There definitely was no liquid level above the bottom tangent line. The gamma scans showed some absorption at the tangent line, about the same absorption as the froth on the trays, but this could have been due to the welds at the tangent line. The neutron scans showed no liquid at all in the reboiler draw compartment.

The absence of liquid level in the reboiler draw compartment gave strong support to the two steady-states theory.

Solution

With the theory verified, the challenge was now shifting from the steady state with the stagnant "A" reboilers back to that of the two reboiler circuits operating simultaneously.

In a circulating thermosiphon reboiler (as distinct from once-through thermosiphon, like the stripper reboiler), increasing the liquid head in the tower generates more liquid circulation through the reboiler circuit and reduces the fractional vaporization[4]. Considering the pressure balance and assuming a constant heat duty, the higher liquid head supplied to the reboiler is matched by the additional friction head due to the enhanced liquid circulation rate plus the additional static head due to higher amount of liquid in the reboiler shell and outlet line. Extending this reasoning to the stripper reboiler, circulating liquid and increasing the liquid head in the reboiler compartment will raise the pressure drop through the active reboiler circuit. This higher pressure drop will induce some of the liquid to follow the easier path (less pressure drop) through the inactive reboiler. Once this liquid gets heated, it will start boiling, and the thermosiphon will be established in the previously-inactive circuit.

One way of generating circulation and increasing the level in the reboiler draw compartment is by raising the level in the bottom draw compartment until it overflows the baffle, and keep the bottom draw compartment full during the circulation. The risk is that the liquid level may ascend to the trays and cause flooding and possible damage there. It is important to avoid excessive buildup of liquid up the tower.

In principle, the level transmitter can be watched during the operation to avoid excessive liquid level. This was difficult in this deethanizer stripper. Since the reboiler return pipe is only two feet above the baffle (Figure 2) and the vapor from the reboiler return blows downwards causing a frothy dispersion, the frothiness may reduce the specific gravity and lead to optimistically low level measurement. Also, the high tap of the level transmitter is just less than two feet above the baffle, which may make it difficult to see when the level rises towards the trays. Fortunately, there is a differential pressure transmitter across the stripper that comes from the upper tap of the level transmitter and therefore can monitor any accumulation of liquid above this.

The plan was to raise the liquid level in the bottom draw compartment to the level of the reboiler return, an increase in height of about 20 feet. When the level was first raised, the level was controlled using the differential pressure transmitter. Subsequently a pair of electronic pressure gauges was added to the reboiler side of the column to monitor the level directly, an additional gauge was added to the product outlet line to monitor the level on the bottom draw side of the baffle. In order to avoid damage due to entrainment of liquid through the trays the dP across the bottom trays was closely monitored.

To raise level on the reboiler side of the baffle, the bottom draw side liquid was overflowed to the reboiler side by reducing the bottom product flowrate. The level rose until it reached the top of the baffle when it 'flat lined' as it overflowed.

After one hour overflowing the baffle, level had increased on the reboiler side by 13 feet. At that time, the following two process indications provided evidence that the "A" side reboilers were starting to circulate liquid:

(1) A rapid decrease in steam consumed by 3B reboiler suggested a decrease in liquid flow to the "B" side reboilers, and

(2) A rapid decrease in 4A hot side outlet temperature suggested an increase in duty in 4A reboiler.

Steam was quickly opened to 3A to ensure we maintained "A" side reboiler circulation. Almost immediately, even split and duties were confirmed between "A" and "B" side reboilers. Bottoms outlet flow was increased to stop overflowing the baffle, and bottoms level was returned to normal.

The thirteen foot level increase was probably not enough to get the level of weathered liquid (the liquid that was sitting in the A reboiler circuit for a lengthy period) up to the reboiler return nozzle on the tower. But the fresh liquid that had entered the reboiler as level was being raised provided enough vaporization to gas-lift the weathered naphtha into the tower and start circulation.

The column remains prone to losing reboiler duty. This is believed to be because a portion of the bottom tray liquid bypasses the reboiler side of the baffle. The bottom tray is an active tray. It remains unclear whether the tray weeps enough liquid to cause the problem or whether some of the liquid directed to the reboiler draw side of the baffle falls into the product side of the bottom. To fix this in the future the bottom tray will be replaced with a total trapout tray to ensure all bottoms liquid must pass through the reboilers.

Until the plant gets to an opportunity to install the chimney tray, the reboiler continues to stall periodically and raising the liquid level has become a standard procedure that has been executed several times since by operations to regain lost reboiler circulation

Lessons Learnt

The investigation highlights the importance of good troubleshooting and problem diagnosis for overcoming bottlenecks. In our case, pressure drop calculations, close design reviews and examination of operating charts, and gamma scans, led to the correct diagnosis that enabled debottlenecking an entire gas plant with no hardware modifications.

Maldistribution between parallel reboiler circuits has received little attention in the literature, with no published case studies picked up by Kister's survey[2]. To the best of our knowledge, this is the first time this type experience is reported. This maldistribution is especially severe at low rates, when the pressure drops in the reboiler circuits are low. Due to the low pressure drop, the fluids find their favourite paths and steady states, and can easily swing from one steady state to another, promoting instability. With parallel reboiler circuits, careful analysis is important of the pressure balance both at the normal and at a reduced rates operation, and should be performed at the design. This case study demonstrates that increasing the pressure drops in these circuits can help route the fluids to the preferred patterns and counter this maldistribution.

Baffled systems have their benefits[1], but can also generate or amplify hydraulic problems, and need to be designed carefully. In the case of the deethanizer stripper, the baffle was far taller than it needed to be. Had it been even a couple of feet shorter, it would have been easy to raise the liquid level in the bottom compartment above it and achieve better operational flexibility.

It is important to provide flexibility when designing once-through thermosiphon circuits. As pointed out[1,2,4], they can be quite troublesome, and simple features that can convert them temporarily to circulating thermosiphon circuits can go a long way towards easing startups and avoiding process bottlenecks.

This case teaches a number of very important key lessons:

- With a parallel reboiler system, maldistribution between parallel circuits does occur and can cause severe bottlenecks.
- The pressure balance both at high and at low rates needs to be considered at the design.
- It is important to provide flexibility in once-through reboiler systems that will allow operating as a circulating thermosiphon system, at least at low rates.
- Finally, good troubleshooting can provide low cost (in this case, no cost) debottleneck of an entire gas plant.

References

1. Fractionation Research Inc. Design Practices Committee "Reboiler Circuits for Tray Columns", Chemical Engineering, p.26, January 2011.
2. Kister, H. Z. , "Distillation Troubleshooting", Wiley-Interscience, NJ, 2006.
3. Kister, H. Z., T. C. Hower Jr., P. R. de M. Freitas, and J. Nery, "Problems and Solutions in Demethanizers with Interreboilers," Proc 8th Ethylene Producers Conf, New Orleans, Louisiana, 1996.
4. Kister, H. Z., "Distillation Operation", McGraw-Hill, NY, 1990.

www.clarksolutions.com.br

Clark Solutions for AIChE's Kister Distillation Symposium 2017

On the Role of Proper Mist Elimination System for Distillation Tower Performance

Bruno B. Ferraro[1], Vitor A. Sturm[1], Michael D. Montani[1], João P. Torgal[1,2], Alexandre C. Bastida[1], Nelson P. Clark[1]

[1]*Clark Solutions, São Paulo, SP, Brazil*
[2]*University of São Paulo, São Paulo, SP, Brazil*

Abstract

Mist eliminators usually represent a low percentage of column internals overall cost, nonetheless these devices can increase tower capacity and separation efficiencies. A proper understanding of mist eliminators fundamentals, design and installation promotes better distillation operating conditions. This paper presents basic operating mechanisms of employed mist eliminators, basic design equations and installation techniques to ensure proper operation of these devices that also play a role in a fine separation process.

Keywords: Mist Elimination, Entrainment, Mist Formation

1 INTRODUCTION

Engineering will always find room for improvement as ideal solutions will never be met. Improvements can be cost-effectively achieved by means of design enhancements, technologies combination or innovations.

Mass transfer applied to distillation is recognized as a thoroughly developed field in regard to technologies and understanding – and its barriers are well known and studied. This article purpose is to advise on the understandings and limitations of one distillation process barrier – entrainment.

The studies for improving capacity or efficiency require the evaluation of basic mechanisms. In distillation, designers and operators research for internals that will produce finer gas bubbles, higher gas-liquid contact at the packing interface and thinner film interfaces, among others.

As a process consequence, entrainment is natural wherever gas and liquid phases are in close contact. As driving forces are further increased inside distillation columns – higher velocities, turbulence and contact, thinner films – to enhance production, an increased entrainment will be onset, and usually finer entrained particles will be generated.

Mist, fine liquid particles laden by a gas stream, is mainly generated by two major mechanisms: mechanical entrainment as a result of combined actions of dynamic shear stress due to forced contact between the liquid and gas phases and solid surfaces with liquid drag by the gas stream, generating droplets of micrometric diameters; and condensation due to thermodynamic changes in the system or chemical reactions in the gas stream, generating droplets of submicron diameters.

Mist eliminators in a properly designed and operated distillation plant will support operation and uptime. A proper selection and design of mist eliminators for an adequate removal of liquid carryover will increase purity and recovery lying in the fact that liquid entrained in process gas causes equilibrium disturbance on downstream stages, hindering proper mass transfer.

2 MECHANICAL ENTRAINMENT

Entrainment is the draw of liquid droplets of varying diameters alongside the gas flow. When two fluids flow together, their interface can become unstable depending on flow velocity and fluid properties such as densities, viscosities and surface tension. This instability may form waves and generate droplets in a wide range of sizes.

For absorption and stripping towers entrainment is the partial transition from countercurrent flow into concurrent flow for a liquid load fraction, partially affecting process efficiency, product composition and recovery.

Inter-tray entrainment is an often discussed subject mainly because it has a direct impact over tray efficiency. Between stages in a distillation there is a higher less-volatile components concentration at the bottom tray. The less-volatile and nonvolatile components can eventually be carried over as entrainment to the tray above and can consequently hinder mass-transfer and affect tray separation efficiency.

Several studies have been conducted over the past decades to determine which operational and design parameters are responsible for entrainment inception, quantity and effects. Colburn (1936) studied the effects of entrainment on tray efficiency. Fair (1961) correlation for crossflow sieve, valve and bubble-cap trays predicts entrainment flooding, including at froth regime for lower tray spacings. Kister and Haas (1987) correlations are good fits for collected data at spray regime for both sieve and valve trays. These studies and correlations have been developed to address not only process conditions but also design parameters such as orifice diameter, tray spacing and clear liquid height. These empirical models rely on large database and do not physically describe entrainment behavior.

After film rupture at the gas-liquid interface, a droplet is propelled at a given angle, rotation and velocity subject to drag, weight and buoyancy. The droplet vertical position will respect the following equation derived from applied position-dependent forces:

$$\rho_L V_L \frac{d^2\vec{x}}{dt^2} = \frac{\rho_G}{2}\left(\frac{d\vec{x}}{dt} - \vec{v_G}\right)^2 C_d A_{L,p} + \rho_G V_L \vec{g} - \rho_L V_L \vec{g} \quad (1)$$

To numerically solve Equation (1) one must know the initial droplet vertical velocity ($\frac{d\vec{x}_0}{dt}$) at the gas-liquid interface ($\vec{x}_0 = \vec{0}$) and properly correlate drag coefficient (C_d). Gas and liquid densities (ρ_G and ρ_L), gas bulk velocity (v_G), droplet volume (V_L) and projected area ($A_{L,p}$) are known. The solution of this equation can show whether a droplet of given diameter (d) is ascending or descending depending on its travelled distance ($\vec{x}$). In other words, this equation determines whether or not a droplet reaches the downstream equipment (or tray above). Froth or liquid jets are not described by Equation (1), which addresses only spray regime.

Drag coefficient studies were reviewed by Aggarwal and Peng (1995).

Holbrook and Baker (1934) studied the effect of slot velocity for different tray spacing in a bubble-cap column. They observed that at lower tray spacing there is a carry-over increase for higher slot velocities whereas the inverse was observed at higher tray spacing. Their results can be partially explained by Equation (1). In both trays spacing at higher slot velocity droplets are assumed to be generated at an averaged higher diameter and quantity because of higher available kinetic energy. Independent to tray spacing, initial entrainment will be the same for the same slot velocity, but accordingly to Equation (1) droplets will tend to decelerate proportionally to diameter (d) as they travel upward ($\vec{x}$) until they either achieve an ascending terminal velocity ($\frac{d^2\vec{x}}{dt^2} = \vec{0}$) and are carried over or they fall back ($\frac{d\vec{x}}{dt} \cdot \vec{v_G} < 0$), reducing entrainment proportionally to droplet distribution as gas rises. A finer distribution would reach higher than a coarser distribution, but a coarser distribution could still be carried over and reach the tray above if spacing is small enough as it was observed by Holbrook and Baker (1934). Therefore, Equation (1) partially explains entrainment stability and disengagement for any defined set of already suspended droplet distribution.

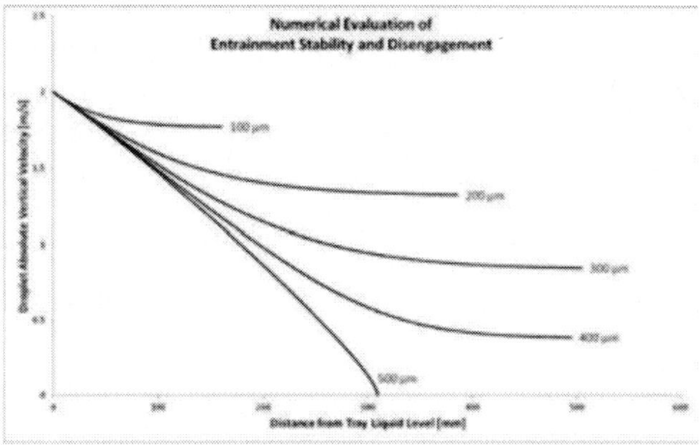

Figure 1. Numerical evaluation of entrainment stability and disengagement (air-water system at 2 m/s).

As vapor flows, gravitational, buoyancy and drag forces acting on droplets tend to equilibrium. The equilibrium point is represented by Equation (1) at $\frac{d^2\vec{x}}{dt^2} = \vec{0}$. A special case is when both $\frac{d^2\vec{x}}{dt^2}$ and $\frac{d\vec{x}}{dt}$ tend to zero.

The Souders-Brown equation (Souders and Brown, 1934) represents this position-dependent force balance special case. Souders-Brown was originally formulated to account for entrainment and capacity on design of fractionating columns. The equation was derived from the force equilibrium between drag (F_d), weight (F_w) and buoyancy (F_b) applied on an averaged droplet diameter and drag coefficient to determine maximum allowed gas bulk velocity in a process:

$$F_d = F_w - F_b \tag{2}$$

Or,

$$\frac{\rho_G}{2}(v_G)^2 C_d A_{L,p} = \rho_L V_L g - \rho_G V_L g \tag{3}$$

Souders-Brown Equation is then stated as:

$$v_G = \sqrt{\frac{4gd}{3C_d}} \cdot \sqrt{\frac{\rho_L - \rho_G}{\rho_G}} \tag{4}$$

And further simplified as:

$$v_G = K\sqrt{\frac{\rho_L - \rho_G}{\rho_G}} \tag{4}$$

The K-factor is often employed to set velocities for sizing a number of processes and internals, among then, trays and mist eliminators.

As a simplification for the de-entrainment problem, the K-factor is widely applied as a pressure dependent constant, relying in an averaged droplet diameter and drag coefficient which does not properly correlate for a wide range of processes and conditions. Notwithstanding its limitations, K-factor is a practical design parameter.

As continuation of Souders and Brown original work, the K-factors, remodeled as Flooding Capacity factors (C_{SB}), are frequently correlated for different operating conditions and design parameters. Equation (5) describes Kister and Haas (1987) correlation:

$$C_{SB} = 0.0277(d_o^2\sigma/\rho_L)^{0.125}(\rho_G/\rho_L)^{0.1}(TS/h_{ct})^{0.5} \qquad (5)$$

While Souders-Brown correlation gives a generalized approach, Kister and Haas (1987) empirical correlation also attempt to implicitly evaluate droplet generation with orifice diameter (d_o), surface tension (σ), liquid density (ρ_L) and vapor to liquid density ratio (ρ_G/ρ_L) as well as implicitly evaluate droplet distribution with tray spacing (TS) and clear liquid height at the froth-to-spray regime (h_{ct}).

Equation (1) only physically describes droplet distribution variation between trays for a given set of droplets.

None of these models properly describe droplet generation initial distribution neither coalescence nor atomization.

3 CONDENSATION

In a distillation column, pressure and temperature gradients applied over a gas stream may cause condensation with respect to composition. Although condensed liquid is consequently a form of liquid entrainment with a different composition compared to the gas stream composition, the gas-condensed liquid stream overall composition that reaches the next stage tray will be the same as the gas stream composition that left the previous stage tray unless there is disengagement. The condensed liquid entrained in the gas stream is composed mainly of less volatile components and would already have been condensed and separated on the next stage tray.

This entrainment is initially made of sub-micron droplets, differently from large droplets observed from liquid carryover. These droplets tend to increase in size depending on pressure and temperature gradient, gas and liquid composition and presence of condensation surface such as liquid flow, existing droplets or vessel wall. High levels of condensation cause internal reflux.

4 MIST ELIMINATION DEVICES

Process engineers currently have a wide range of mist elimination equipment to select depending on generated mist particle size distribution.

The simplest form of current mist eliminators employed in industry are sets of crimped parallel plates which cause gas direction change along channels, capturing particles by inertial impaction at every deflection. Chevron or vane mist eliminators parallel plates are separated by distances that typically vary from 10 to 50 mm with 3 to 6 passes.

Chevron mist eliminators bring lower efficiency than other mist eliminators, but also lower pressure drop and higher liquid draining capacity. Chevrons are usually the best option for high fowling operation as blades support higher drainage and allow simple washing when plugged.

Figure 2. MaxiChevron® mist eliminator

Clark Solutions MaxiChevron® Double-Pocket™ is similar to a conventional chevron. However, a set of pair of blades with draining slots replaces conventional blades, promoting draining. Therefore, this equipment is able to operate at even higher vapor capacities than those achieved by conventional chevron in the same service at the expense of higher pressure drop and lower fowling capacity.

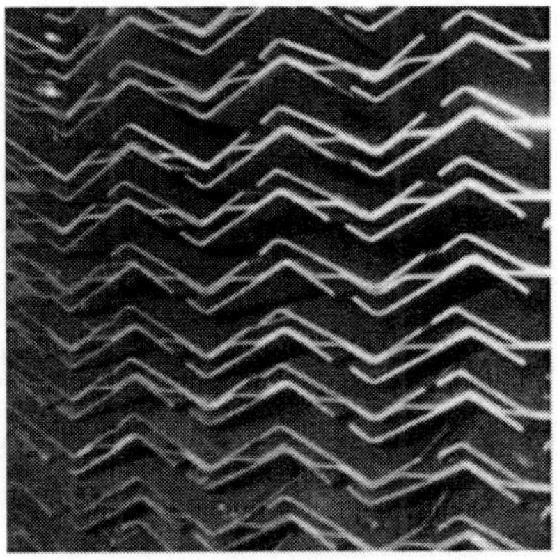

Figure 3. Clark Solutions MaxiChevron® Double-Pocket™

For higher efficiency, fine metal wire provides much smaller and difficult to avoid obstacles than large metal sheet plates. Wire mesh pads are sets of knitted wire mesh, arranged in particular geometries to capture small particles.

Figure 4. MaxiMesh® mist eliminator

Originally developed in the early 1950's with few different mesh densities, wire mesh pads have developed substantially in the last 70 years. Today, they can be offered in a wide variety of geometries, wire sizes and knitting with or without polymers which will allow capture of finer particles in the micrometric range.

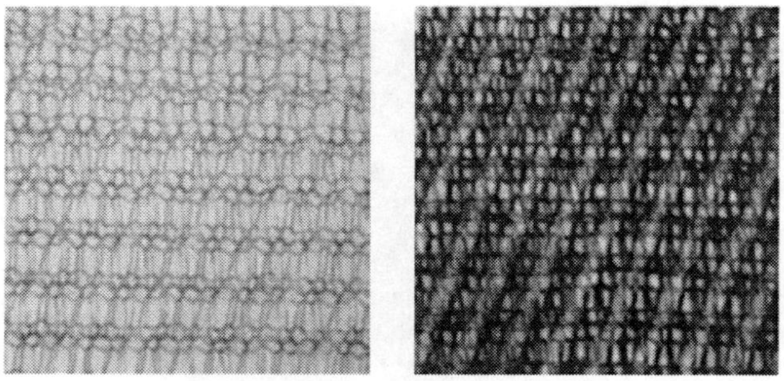

Figure 5. Knitted and Co-knitted mesh

Plain metal wire meshes are designed to achieve capture of particles with few micrometers. More efficient designs will generally require a co-knit mesh, but will also increase pressure drop. Special applications may require special alloys or plastics for the wire mesh structure.

There are other types of mist eliminators such as candle filters and swirl flow cyclones. However, chevron and wire mesh mist eliminators represent most applications in distillation. This paper discusses wire mesh mist eliminators because of its broadest use.

5 DE-ENTRAINMENT THEORY

De-entrainment is the disengagement of suspended droplets within gas flow. Equation (1) describes de-entrainment as an inflexion of the droplet vertical velocity vector in respect to gas flow ($\frac{d\vec{x}}{dt} \cdot \vec{v_G} < 0$). But de-entrainment can be further increased by mist elimination collection mechanisms.

The drawback of mechanically achieving de-entrainment is pressure drop. Studies have been conducted to describe and correlate collection mechanism efficiency, overall efficiency and pressure drop from dry to flood operations.

Different devices exist in order to separate the entrained liquid from the gas phase, making use of three major mechanisms to separate and coalesce droplets.

Inertial impaction occurs when a gas stream carrying liquid particles passes through an obstacle. The gas will avoid the obstacle, but particles with sufficient inertia will be projected from the gas flow and impact on the obstacle. Inertial Impaction will be as pronounced as the gas velocity raises, as liquid to gas density ratio increases and as target gets smaller (*i.e.* as steep is the gas trajectory change).

Direct interception occurs when particle diameter is small enough to follow gas streamlines, but big enough to collide on obstacles' edges. This effect will increase as the target gets smaller and the mist eliminator void fraction is reduced.

Brownian diffusion occurs with extremely small particles which acquire a Brownian movement caused by energy transfer on collision with randomly moving gas molecules. This movement will drive these particles to eventually collide with the mist eliminator targets. Brownian diffusion capture will be favorably affected by the reduction in target diameter and bed void fraction. Lower gas velocities will increase particle residence time within the mist eliminator bed, resulting in a higher probability collection, increasing the overall efficiency of the mist eliminator designed for this latter capture mechanism

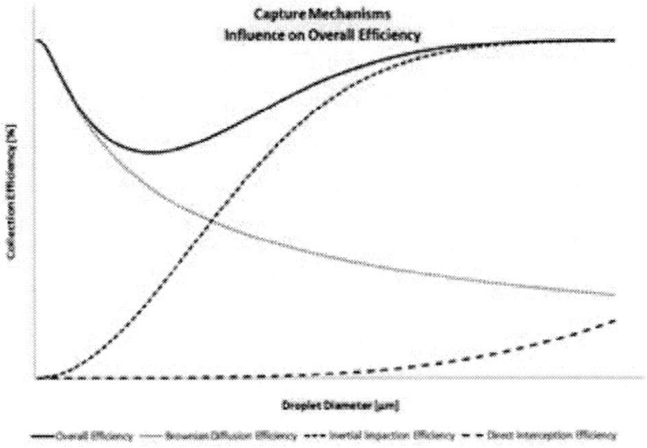

Figure 6. Collection efficiency for different capture mechanisms

(Co-)Knitted wire mesh pads mist eliminator devices collects droplets primarily via inertial impaction, thus, the droplet size range collected is specific. Since entrainment raised from mechanically dragged droplets in distillation towers contains size distributions fit for this equipment efficiency, wire mesh pads are the first de-entrainment devices to be considered – unless fowling is a primary concern.

The efficiency of a filtering media can be expressed as the fraction of the particle distribution that is collected. One way to account for efficiency is to sum contributions of individual wires or fibers, assuming that they are equal in size and provide the same individual collection efficiency for the uniformly packed bed. This can be expressed as:

$$E = 1 - exp\left[-f\sum \eta\right] \qquad (6)$$

The above equation is derived originally from the Kleinschmidt equation (Kleinschmidt, 1939) designated to calculate particle collection efficiency of Pease-Anthony cyclonic scrubbers, and was later used for particle collection on fiber filters, where the exponential term is also referred as Penetration (Pan and Zhong, 2006). The equation was formulated to account for single filament efficiency and generalized for the equipment extension.

This equation (6) can be used for wire mesh mist eliminators, with f representing the effective contact area expressed as $\frac{-A_e LF}{\pi}$. Mesh pad thickness (L), specific surface area (A_e) and fraction of wires in an adequate collection position (F) are usually known.

The efficiency sum in equation (6) can be stated as the contribution of inertial impaction (η_{II}), direct interception (η_{DI}) and Brownian diffusion (η_{BD}) minus the interfering efficiency of interacting mechanisms (η_{IM}):

$$\sum \eta = \eta_{II} + \eta_{DI} + \eta_{BD} + \eta_{IM} \qquad (7)$$

For wire mesh pads, the most substantial contribution for equation (7) is from inertial impaction.

This approach, although a good method, may sometimes be misleading for some mesh-type mist eliminator devices since specific geometric parameters (*e.g.* layering of the knitted meshes and crimps) are not taken in consideration, oversimplifying the mesh pads into a bed of wires.

Therefore efficiency models must be evaluated before applying on wire mesh mist eliminators, since some equation premises are not always met. Mead-Hunter et al. (2014) reviewed several efficiency models from literature. One widely used single target efficiency model was developed by Langmuir and Blodgett (1946).

More specific equations were developed for wire mesh mist eliminators overall equipment efficiency, such as the semi-empirical model of Carpenter and Othmer (1955) and the mechanistic models of Brunazzi and Paglianti (1998 and 2000). These models rely on single target efficiency models such as Langmuir and Blodgett (1946) to predict overall equipment efficiency along all wire mesh layers. Brownian diffusion and direct interception can be neglected for the evaluation of wire mesh mist eliminators efficiency as shown by Holmes and Chen (1984).

Both Carpenter and Othmer (1955) semi-empirical model and Brunazzi and Paglianti (1998) assume no re-entrainment and no liquid hold-up. While Carpenter and Othmer (1955) consider droplets remixing along wire mesh layers, Brunazzi and Paglianti (1998) consider no mixing in their multi-cell efficiency model.

Operating outside creeping flow droplets may not experience all cells as assumed by Brunazzi and Paglianti (1998). Droplets may be disturbed and eventually follow interstitial paths between cells, which can explain Brunazzi and Paglianti (1998) overestimated efficiency.

Most correlations for single fiber efficiency through different mechanisms assumes specific flow fields that exist when diameter-based Reynolds numbers are very low, such as Stokes or Kuwabara Flow Fields (Kuwabara, 1959). For typical vapor velocities the flow field around mesh wires may differ significantly from creeping flow. For these reasons, many wire mesh applications may be misrepresented.

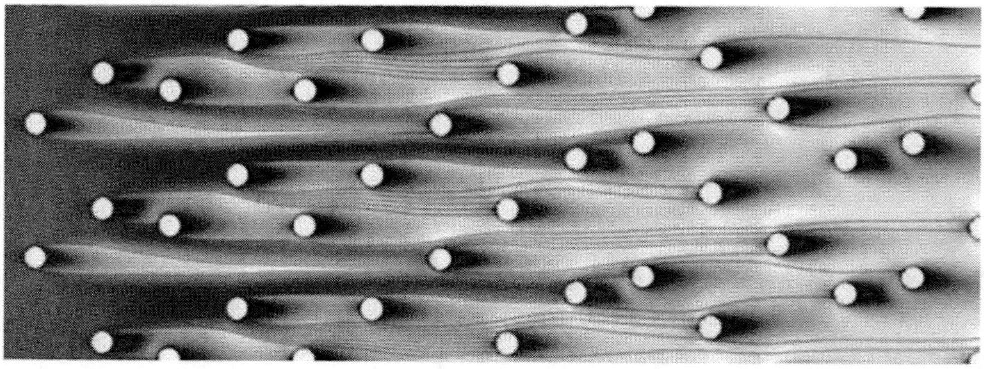

Figure 7. Computational Fluid Dynamics for wire arrays (approximation velocity = 2 m/s, gas density = 1.2 kg/m³, particle density = 1.000 kg/m³, wire diameters = 0.3 mm, particle diameter = 10 µm, void fraction =0.96) – layers are not always aligned, but wire spacing inside layers are the same

When designing wire mesh mist eliminators, efficiency and pressure drop balance is sought. Various factors influence equipment pressure drop (*e.g.* operational conditions, mesh thickness, mesh geometry, wire diameters and void fraction). Generally, higher efficiencies bring higher pressure drop. Co-knitting allows for a higher increase in efficiency to pressure drop ratio. Multi-zones allow increase in capacity as lower zones can be designed with lower density mesh and upper zones with co-knitted higher density mesh.

Figure 8. Clark Solutions MaxiMesh® with dual zones (plain metal bottom drainage zone and fiberglass co-knitted high efficiency zone)

Efficiency and pressure drop are often correlated under specific conditions, processes and equipment styles. Many of these correlations are proprietary models.

El-Dessouky et al. (1999) presents a set of correlations for capacity, efficiency and pressure drop built from a series of experiments in typical mesh pads specifically used in multi-stage flash desalination units. Al-Dughaither (2010) studies the behavior of mesh pads on low velocity bubble columns.

6 PRACTICAL DESIGN

Although technology develops and more products and tools become available, many designs still follow previous standards. Souders-Brown equation and empirically determined K-factors are a dominant design technique.

On recurrent cases, and when pressures are near atmospheric, simple standard equipment such as plain metal wire mesh pads and single pocket chevrons can be calculated using Souders-Brown equation with known K-factor values.

As an example, a Maximesh® Style 431 wire mesh pad for an air and water system with an actual gas flow of 10,000 m³/h, operating at 1 bar and 25°C, can have its ideal area calculated following Souders-Brown equation as:

$$K_{431,1\,bar} = 0.106[m/s]$$

$$V \cong 0.106[m/s]\sqrt{\frac{997[kg/m^3] - 1.2[kg/m^3]}{1.2[kg/m^3]}} \cong 3.0[m/s]$$

$$A \cong \frac{10{,}000[m^3/s]}{3.0[m/s]} \cong 0.9[m^2]$$

Clearly, K-factors should be corrected for high pressures and more precise correlations could retrieve a more compact equipment area. Pressure changes, substantial gas or liquid density variations, excessive liquid loads and other factors can lead to a different choice of mist eliminator design and area.

This ideally calculated area may be modified through more detailed calculations to achieve desired droplet diameter range removal efficiency. Nonetheless, this is a valuable widely used method for estimating mist eliminator sizes.

7 INSTALLATION

Proper equipment installation is a major requirement for mist eliminator system performance.

Wire mesh and chevron mist eliminators are designed to fit normal sized manholes of 600 mm (24"). Also, sections should not weight more than 20-25 kg to allow for a single man to hold and manipulate the individual sections. As a general rule, sections width will be around 350-500 mm.

Depending on tower diameter the number of sections will change. They can be full length, half length or even cut in thirds, fourths or sixties, depending on sections weight and tower diameter, similarly to structured packing and trays.

Mist eliminator arrangement and flow direction is chosen depending on calculate area, space availability and geometry as well as operating conditions.

Inside distillation, absorption or stripping towers, mist eliminators installation are mainly horizontal as gas vertically flows.

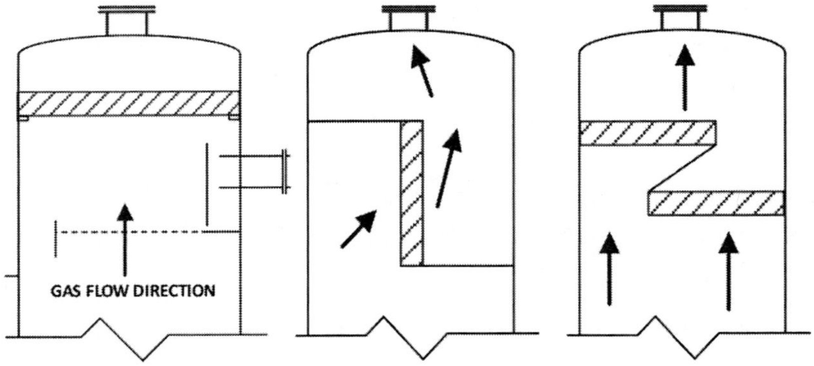

Figure 9. Mist eliminators possible arrangements

However, other arrangements are possible, such as vertical installation with horizontal flow and Z-shapes, depending on design requirements. Computational Fluid Dynamics can assist on design evaluation and improvement of mist eliminators geometry and installation, assessing mist eliminators effective area in respect to flow velocity fields.

For applications where operating pressures are close to atmospheric, some rules of thumb can be applied regarding mist eliminator location.

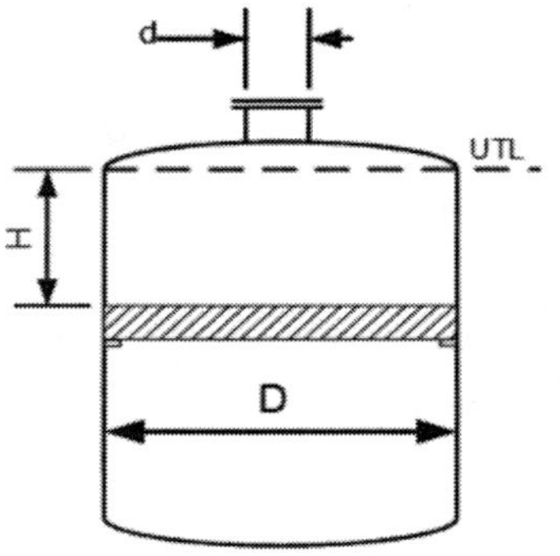

$$H \geq \frac{D-d}{2}$$

$$H \geq \frac{D+d}{2}$$

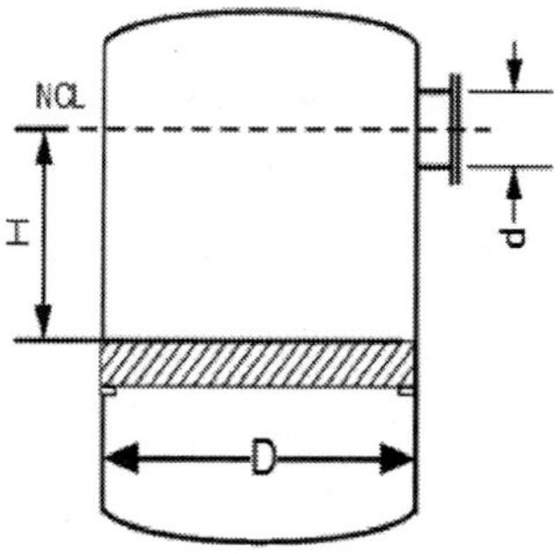

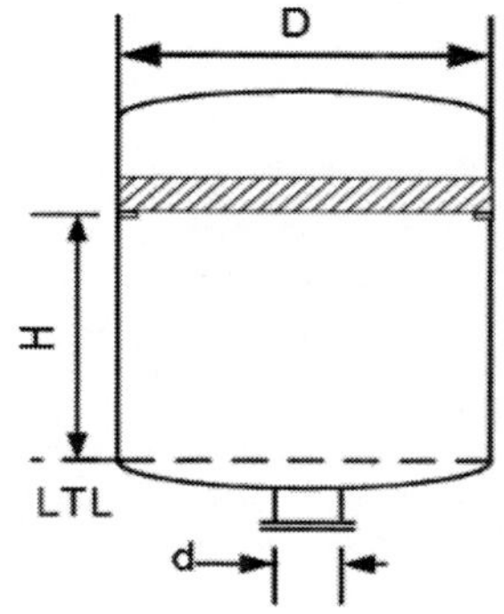

$$H \geq \frac{D}{2} - \frac{d}{3}$$

$$H \geq \frac{D + d}{2}$$

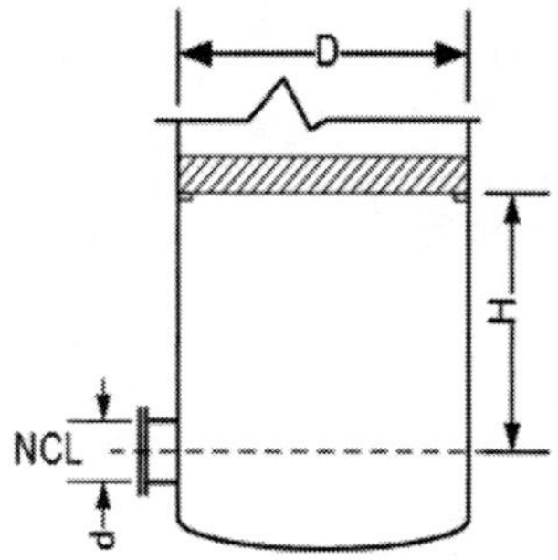

Figure 10. Mist eliminators positioning practical standards

Simulations can be made to evaluate and achieve more compact vessels. Proper installation is sought to enhance gas distribution across mist eliminator surface as uneven flow distribution will result in uneven velocity distribution and compromise performance.

There is a wide range of fastening options to install mesh pads. Tie wires fasten mist eliminator sections to a support ring welded or attached to the vessel wall. The ring contains holes that are usually spaced by 200-250 mm, depending on mist eliminators sections width.

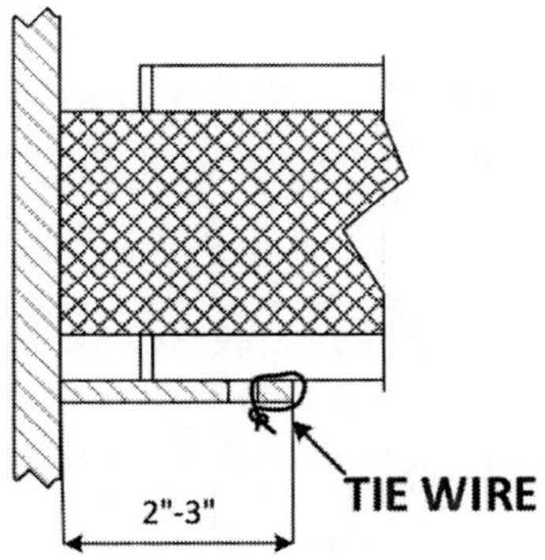

Figure 11. Tie wire fastening

Tie wires can be applied provided a non-corrosive and low pressure service. High pressure systems may submit tie wires to stresses that eventually result one or more sections displacement.

On corrosive or more demanding applications, J-Bolts are the preferred choice. J-Bolts offer a more resistant and stronger fastening between the mist eliminator and the support ring.

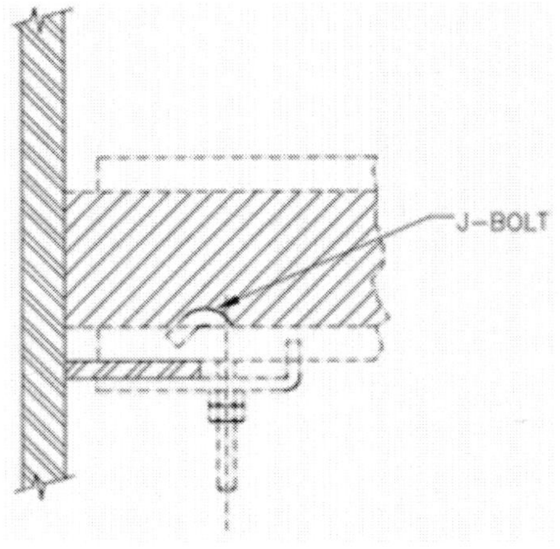

Figure 12. J-bolt fastening

If bottom access is difficult, a third option is the L-Bolt. Similarly to a key, once the section is in place, turning the bolt to the proper position will firmly fasten the section.

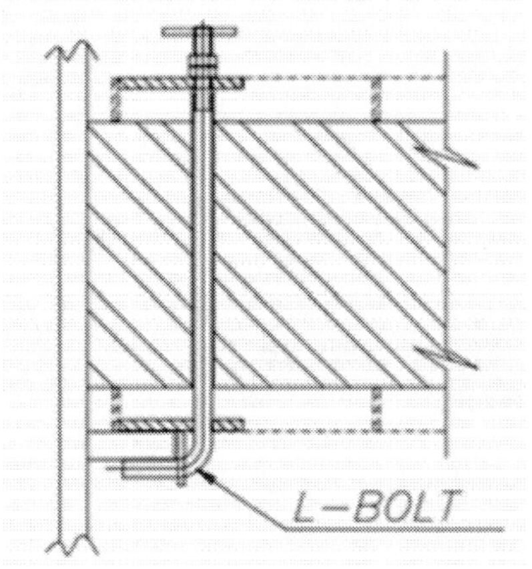

Figure 13. L-bolt fastening

There are other ways to install mist eliminators (*e.g.* between rings with removable sections or using upper support bars). Selection is usually made based on operating conditions and maintenance comfort.

8 APPLICATIONS

Typically the rectifying section is above the feed tray, where top vapor is corrected to achieve desired composition. On this section both foaming and entrainment are less significant due to lower vapor density and velocity. As such, tray spacing can be reduced at the rectifying section with lower compromise of system capacity.

The stripping section, on the other hand, experiences higher vapor density and velocity and as a consequence an increased foaming and entrainment. Solids, when present, are also mainly observed on the stripping section. Therefore, higher tray spacing may be used to guarantee system capacity and avoid mixing between trays. An alternative solution is to provide entrainment removal between trays. Some applications, such as a whiskey still, utilizes an entrainment removal section only between the stripping and rectifying sections, consisting solely of a plate.

But for some operations the feed tray is above the rectifying section, such as the wine feed on sugarcane alcohol distillation. Therefore both liquid and solid entrainment must be removed from the phlegm lateral removal below feed. In some cases a wave-plate mist eliminator is positioned between the distillation column and the rectification column in order to prevent entrainment to reach the rectification column feed.

There are plenty of other well proven typical mist eliminators applications such as wire mesh pads installed at the gas exit of overhead condensers to protect downstream compressors and increase condenser's throughput; chevron or fast draining wire mesh pads such as Maximesh® Style 353 installed on vacuum tower's steam stripper section; and very high efficiency, co-knitted wire mesh pads installed on glycol, amine and solvent drying and absorption towers to reduce carryover losses.

As a practical example one can consider a Jet-A side stream stripper producing 15,000 BPD of Jet-A. Assuming a 0.5% carryover from the stripper the amount of Jet-A which was carried to the Naphtha section would be 75 BPD of Jet-A, roughly $ 125,000 in losses per year considering a $ 5.00 price difference per barrel between Jet-A and Naphtha. A wire mesh pad of 5 square meters for this service will cost around $ 5,000. Payback will therefore be less than 30 days.

This same approach can be applied for other applications. While the economics may not be as favorable, payback is typically lower than 3 to 4 months.

9 CONCLUSION

The growth in mist elimination science reflects the perceived interest in attainable product recovery and purity with these low investment equipment. Proper selection and design of mist eliminators are a cost-effective approach to increase operational revenue and may support capacity increase.

Although entrainment is known as a distillation barrier for nearly a century, mist elimination is still a developing science. New studies and modern computational tools offer new perspectives in the understanding of mist eliminators as new products and developments are established.

10 REFERENCES

Aggarwal, S. K. and Peng, F. 1995. A Review of Droplet Dynamics and Vaporization Modeling for Engineering Calculations. Journal of Engineering for Gas Turbines and Power 117: 453-461.

Al-Dughaither, A. S.; Ibrahim, A. A. and Al-Masry W. A. 2010. Investigating droplet separation efficiency in wire-mesh mist eliminators in bubble column. Journal of Saudi Chemical Society 14(4): 331-339.

Al-Sheikh, J. N. 1959. A Study of Entrainment in Distillation Column for Sieve Plate. Master Thesis, Rice University, Houston. 59 p.

Bujake, J. E. 2007. Beverage Spirits, Distilled. Kirk-Othmer Encyclopedia of Chemical Technology.

Brunazzi, E. and Paglianti, A. 1998. Design of wire mesh mist eliminators. American Institute of Chemical Engineers Journal 44(3): 505.

Brunazzi, E. and Paglianti, A. 2000. Design of complex wire-mesh mist eliminators. American Institute of Chemical Engineers Journal 46(6): 1131-1137.

Carpenter, C. L.R. and Othmer, D. F. 1955. Entrainment removal by a wire-mesh separator. American Institute of Chemical Engineers Journal 1(4): 549-557.

Colburn, A. P. 1936. Effect of Entrainment on Plate Efficiency in Distillation. Industrial and Engineering Chemistry. 28(5): 526-530.

Cristini, V. and Renardy, Y. 2006. Scalings for Droplet Sizes in Shear-Driven Breakup: Non-Microfluidic Ways to Monodisperse Emulsions. FDMP 2(2): 77-93.

El-Dessouky, H. T.; Alatiqi, I. M.; Ettouney. H. M. and Al-Deffeeri, N. S. 2000. Performance of wire mesh mist eliminator. Chemical Engineering and Processing: Process Intensification 39(2): 129-139.

Fair, J. R. 1961. How to predict sieve tray entrainment and flooding. Petro/Chem. Eng. 33(10): 45-52.

Holbrook, G. E. and Baker, E. M. 1934. Entrainment in a Bubble-Cap Distillation Column. Industrial and Engineering Chemistry 26(10): 1063-1067.

Holmes, T. L. and Chen, G. K. 1984. Design and selection of spray/mist elimination equipment. Chemical Engineering 91(21): 82-89.

Ishii, M. and Grolmes, M. A. 1975. Inception Criteria for Droplet Entrainment Two-Phase Concurrent Film Flow. AIChE Journal 21(2): 308-318.

Kister, H. Z. and Haas, J. R. 1987. Sieve tray entrainment prediction in the spray regime. Institution of Chemical Engineers Symposium Series No. 104.

Kister, H. Z. and Haas, J. R. 1990. Predict Entrainment Flooding on Sieve and Valve Trays. Chemical Engineering Progress 86(9): 63-69.

Kleinschmidt, R. V. 1939. Factors in spray scrubber design. Chem. Met. Eng. 46: 487.

Kuwabara, S. 1959. The forces experienced by randomly distributed parallel circular cylinders or spheres in a viscous flow at small Reynolds numbers. Journal of the Physical Society of Japan 14(4): 527-532.

Langmuir, I. and Blodgett, K. 1946. A mathematical investigation of water droplets trajectories. Collected works of Irving Langmuir. Pergamon Press 10: 348-393.

Pan, N. and Zhong, W. 2006. Fluid transport phenomena in fibrous materials. Textile Progress 38(2): 1-93.

Pyott, W. T.; Jackson, C. A. and Huntington, R. L. 1935. Factors Affecting Entrainment in Bubble-Cap Columns. Industrial and Engineering Chemistry 27(7): 821-825.

Souders Jr., M. and Brown, G. G. 1934. Design of Fractionating Columns: I. Entrainment and Capacity. Industrial & Engineering Chemistry 26(1): 98-103.

Ujile, A. A. and Iminabo, J. 2014. Evaluating Sieve Tray Flooding in a Distillation Column Using Kister and Haas; and Fair's Correlations. Chemical and Progress Engineering Research 25: 16-23.

485321 Troubleshooting Tower Carryover

Abstract:

For most traditional TEG contactors, high capacity structured packing is the ideal mass transfer device due to the low liquid rate in this service. However, structured packing should only be used in a fairly clean system as it is susceptible to plugging. For dirty systems, bubble cap trays have been traditionally applied. There are a few high capacity tray technologies available that can be used to achieve a capacity boost over bubble cap trays albeit with higher glycol carryover. These new high capacity mass transfer devices exceed the capacity of most existing de-entrainment equipment, such as wire mesh and vane packs. The capacity of demisting devices is further reduced with increased entrained liquid loading as the high capacity devices push to higher superficial velocities. Thus, TEG carryover can ultimately dictate the capacity of the TEG contactor if the de-entrainment internals are not specified correctly or not modified during a revamp to handle the additional TEG carryover. High concentrations of H_2S and CO_2 and even small amounts of contaminants such as heavy hydrocarbon liquids, iron sulfides, salts and other organics in the feed gas can also present operational difficulties. These contaminants can be absorbed in the TEG, accumulate, foul, foam and react to form other byproducts. Therefore, the cleanliness of the system must be considered when selecting internals.

This article summarizes the lessons learned in debottlenecking a "dirty" TEG contactor at our LaBarge Black Canyon Facility in Wyoming US. The facility modifications included replacement of all bubble cap trays with fixed valve trays, modifications to the downcomer inlet sump and seal pan, addition of a new Inlet Vane Diffuser at the gas inlet to the tower, a new full stream carbon filtration system, and optimization of the mist eliminator section at the top of the tower. This work was performed in progressive steps between 2001 and 2004. After completion of this work, the following conclusions can be drawn:

- A clean TEG system will minimize foaming and increase capacity as well as decrease maintenance, operating expense and downtime. As demonstrated, the full stream carbon filtration system reduced TEG liquid carryover by 65-70%.
- Although vendor data was available on demisting cyclones with glycol at elevated pressure and good experience was also available with cyclones in hydrocarbon service at high pressure, this application was outside the data range. The entrained liquid load is also an important parameter that often is ignored or is unknown.
- The combination of the high capacity trays and mist elimination equipment, along with the new solvent filtration system, allowed the gas capacity of the Black Canyon Facility to be almost doubled.

479773 A Better(?) Way to Calculate the Slope of the Equilibrium Line

Abstract:

The calculation of tray efficiencies or the HETP of packed columns requires, among other things, the slope of the equilibrium line. For binary systems the slope is unambiguously defined (although not always rigorously calculated). However, there is no easy way to use a definition of the slope for multicomponent systems because the vapor liquid equilibrium is represented by a surface (in hyperspace if there are more than 3 compounds).

Thus, for multicomponent systems any pseudo-binary method for estimating tray efficiencies and HETP must rely, in part, on an approximate method of estimating the slope. Particularly, information regarding thermodynamic interactions is lost using pseudo-binaries and the slope of the equilibrium line may, therefore, have very limited meaning for a multicomponent system.

In a presentation given at the 2015 National Spring Meeting of the AIChE we looked at four different methods for approximating the slope of the equilibrium line for multicomponent systems. It was shown that these different methods could lead to dramatically different estimates of the equilibrium slope.

Our purpose here is to take further our earlier work and propose a different and, we believe, definitive way to approximate the slope.

The method requires us to define a design component, one per section of the column. The design component can be defined as that which is to be removed from a designated column section. For example: in a distillation to separate the four component mixture ABCD into the AB and CD fractions then the design component for the top section is C and the design component section for the bottom section is B. (In this example these design components are like the heavy and light keys respectively, but that is not always the case, e.g. for column configurations involving side streams.)

The slope then is estimated from two similar flash calculations in which only the design component has its molar amount altered (slightly). This methodology allows proper consideration of thermodynamic interactions and thereby avoids the short comings of adopting a pseudo-binary approach.

We will show that for binary systems this method is exactly equivalent to the rigorous definition of the slope. We will further demonstrate that this method provides a sensible approximation in cases when other published methods have the potential to lead one astray.

478892 Novel Optimization of a Dividing Wall Distillation Column

Abstract:

Novel Optimization of a Dividing Wall Distillation Column

Bailee Roach, Bruce Eldridge

James R. Fair Process Science and Technology Center

University of Texas at Austin

Originally patented in 1949, the dividing wall distillation column has gained renewed attention due to the potential for reduced energy consumption, an estimated thirty to fifty percent over traditional multiple column arrangements, and lower capital investment. Unfortunately, this new interest has not been accompanied by the development of rigorous design approaches that utilize the latest in computational methods. In the current study, optimization and process simulation software was seamlessly integrated to yield minimum energy consuming dividing wall column (DWC) designs over a range of operating conditions and product purity targets. The modeling effort was supported by data collected from the Separations Research Program (SRP) experimental column. Multiple chemical systems were investigated with the study yielding a powerful methodology for future DWC retrofits or new column construction.

2017 AIChE®
SPRING MEETING & 13TH GLOBAL CONGRESS ON PROCESS SAFETY

March 26–30, 2017
Henry B. Gonzalez Convention Center
San Antonio, TX

17 AIChE Spring Meeting, San Antonio, TX — 13TH GLOBAL CONGRESS ON PROCESS SAFETY

17 AIChE Spring Meeting, San Antonio, TX — 13TH GLOBAL CONGRESS ON PROCESS SAFETY

Improve Energy Efficiency of DWCs

Manish Bhargava
Licensing Manager
GTC Technology

GTC Technology
Engineered to Innovate®

Norihito Suzuki
Staff Engineer
TonenGeneral Sekiyu K.K.

TonenGeneral

Concept of Dividing Wall Columns

Conventional 2-Column Fractionation Sequence For Separating Three Components

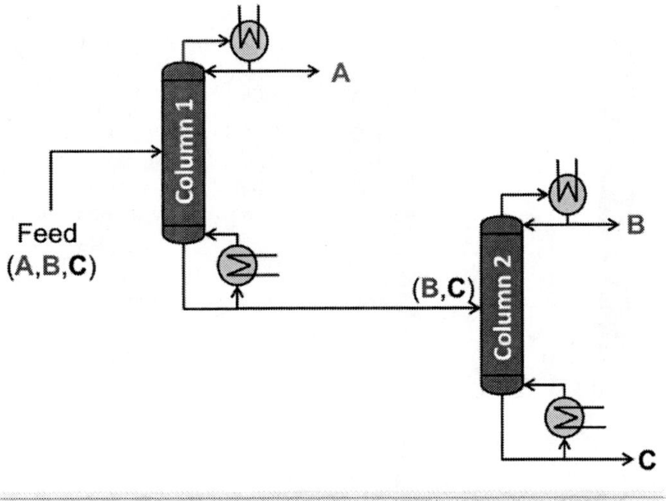

Thermodynamics Inefficiencies in Column 1

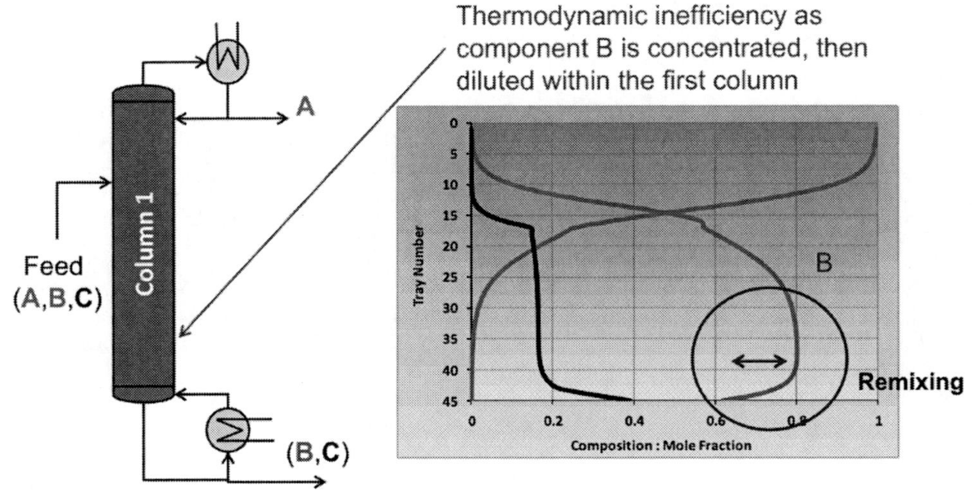

Pre-fractionator Main Column Arrangement

Composition Profile

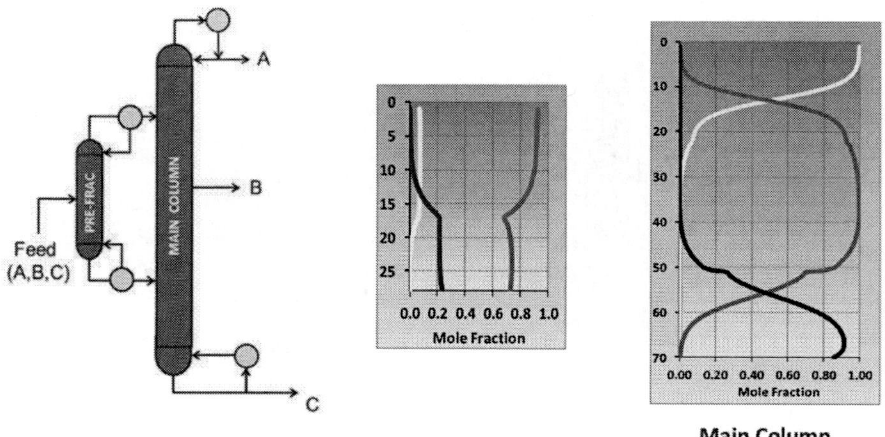

Pre-fractionator Column **Main Column**

2017 AIChE® SPRING MEETING & 13TH GLOBAL CONGRESS ON PROCESS SAFETY

17AIChE Spring Meeting, San Antonio, TX — 13TH GLOBAL CONGRESS ON PROCESS SAFETY — GTC Technology · Engineered to Innovate®

From Prefractionator Arrangement To Dividing Wall Column

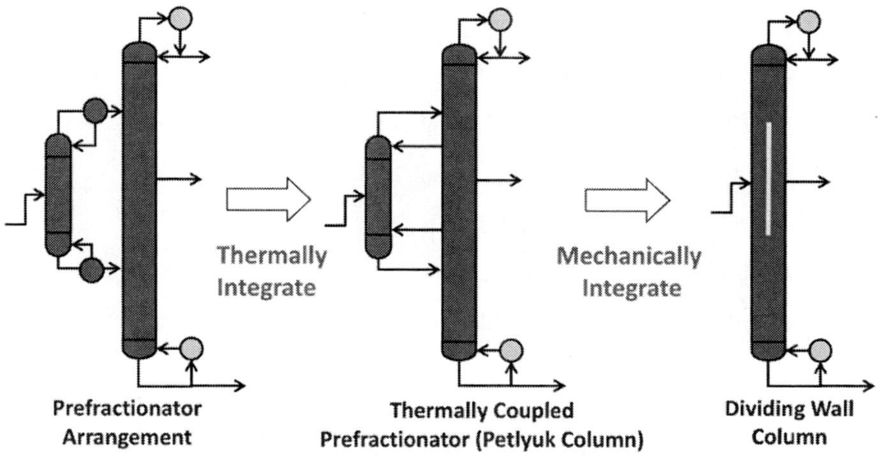

2017 AIChE® SPRING MEETING & 13TH GLOBAL CONGRESS ON PROCESS SAFETY

17AIChE Spring Meeting, San Antonio, TX — 13TH GLOBAL CONGRESS ON PROCESS SAFETY — GTC Technology · Engineered to Innovate®

Typical Benefits of GT-DWC℠ (Dividing Wall Columns)

GT-DWC℠ CAPEX is lower by 20-30%

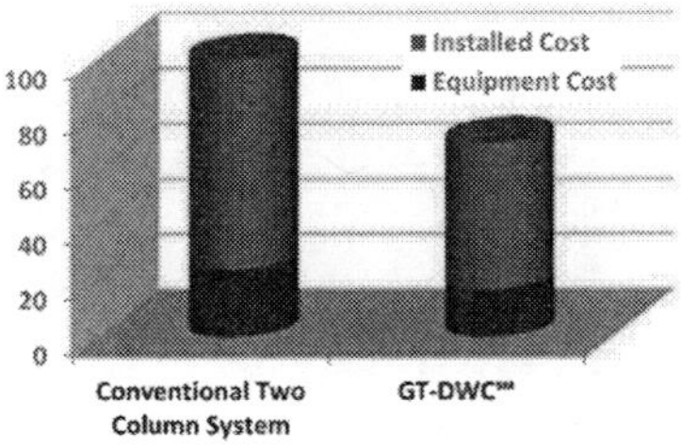

GT-DWC℠ OPEX is lower by 20-30%

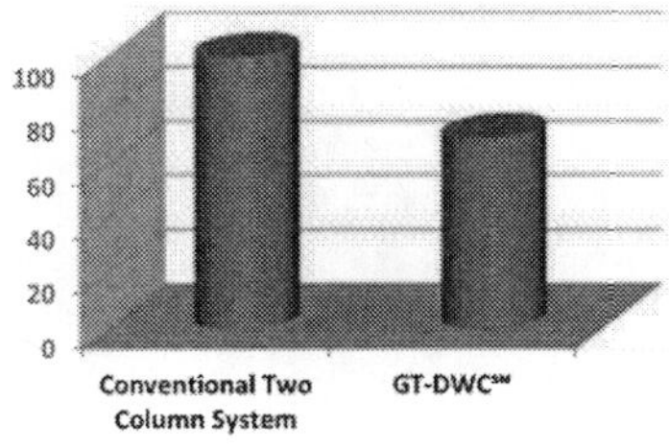

Equipment Count & Plot Space

	Conventional Two Column System	GT-DWC℠
Column	2	1
Reboiler	2	1
Condenser	2	1
Pump	4	3
Plot Space	100%	Approximately 30% less

2017 AIChE® SPRING MEETING & 13TH GLOBAL CONGRESS ON PROCESS SAFETY

17 AIChE Spring Meeting, San Antonio, TX
13TH GLOBAL CONGRESS ON PROCESS SAFETY
GTC Technology - Engineered to Innovate®
Toner General

Typical Dividing Wall Column Applications

- Naphtha Splitters
- Benzene/Toluene/Xylene/C9 Separation
- Debutanizer/Depropanizer
- Depropanizer/Deethanizer
- Reformate Splitters
- Naphtha Splitter/Deisohexanizer
- Dependanizer/Naphtha Splitter

2017 AIChE® SPRING MEETING & 13TH GLOBAL CONGRESS ON PROCESS SAFETY

17 AIChE Spring Meeting, San Antonio, TX
13TH GLOBAL CONGRESS ON PROCESS SAFETY
GTC Technology - Engineered to Innovate®
Toner General

Case Study 1 : Confidential Client in India
Two-column configuration modified to a single column

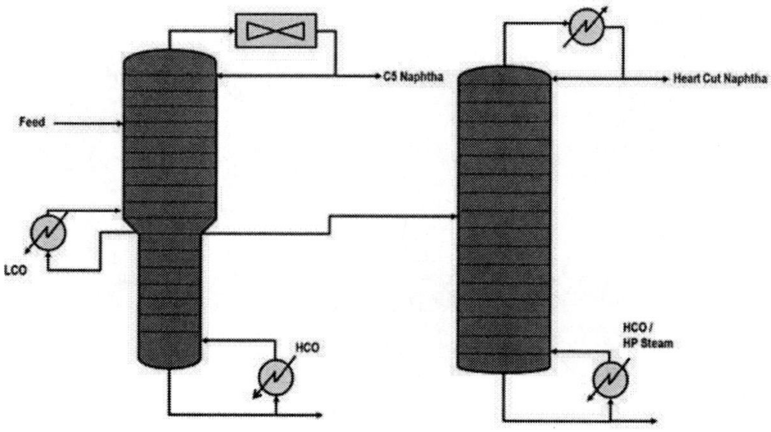

2017 AIChE® SPRING MEETING & 13TH GLOBAL CONGRESS ON PROCESS SAFETY

Objectives for Revamp

- Reduce utility consumption for the separation.
- Produce heart cut naphtha in Naphtha Splitter-1 column, while removing Naphtha Splitter-2 from service.

Bottlenecks in Column Design

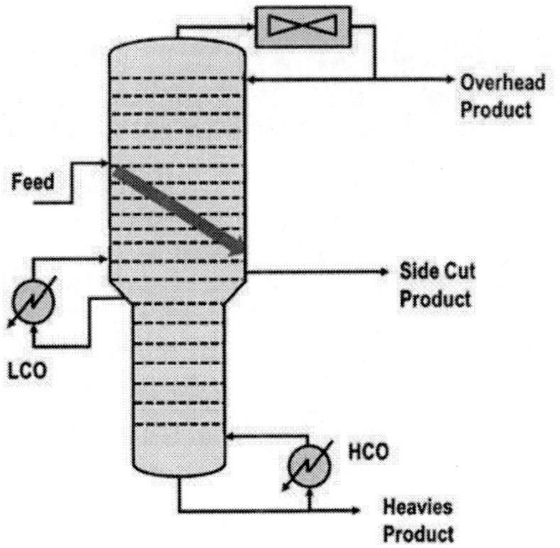

Intermixing of Feed with Side Cut

Lower Diameter Section Puts a restriction on Vapor Liquid Traffic

Revamp to GT-DWC℠

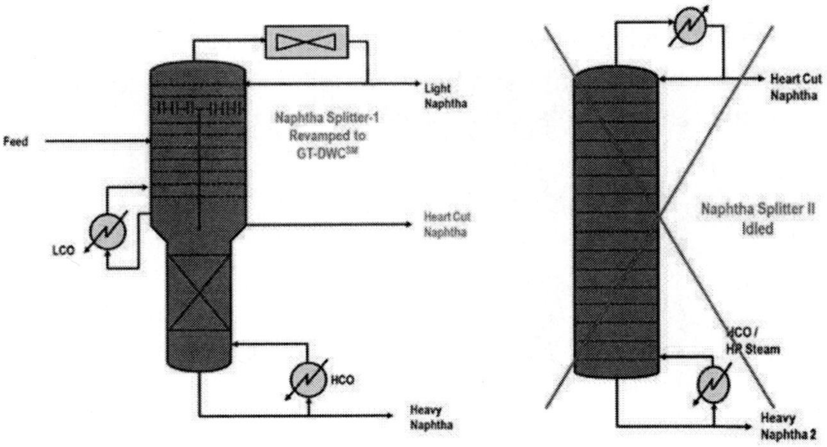

17 AIChE Spring Meeting, San Antonio, TX
13TH GLOBAL CONGRESS ON PROCESS SAFETY
GTC Technology Engineered to Innovate® TonerGeneral

Project Economics

Parameter	Units	Original Design (NS-I and NS-II)	Naphtha Splitter– I after Revamp to GT-DWC℠
Feed	TPH	308.0	308.0
Side Draw Flow Rate	TPH	165.0	165.0
D86 (IBP-FBP)	° C	110.6-170.0	110.0-170.0
Total Heating Duty	MMkcal/hr.	38.0	28.0

2017 AIChE® SPRING MEETING & 13TH GLOBAL CONGRESS ON PROCESS SAFETY

Benefits of GT-DWC℠

- Heating duty reduced by 26%.
- HP steam usage eliminated.
- Naphtha Splitter-II is idled.
- Heart Cut Naphtha obtained in Naphtha Splitter-I.

17AIChE Spring Meeting, San Antonio, TX
13TH GLOBAL CONGRESS ON PROCESS SAFETY
GTC Technology · Engineered to Innovate® TonenGeneral

Dividing Wall Column Trays

2017 AIChE® SPRING MEETING & 13TH GLOBAL CONGRESS ON PROCESS SAFETY

Top View of Dividing Wall

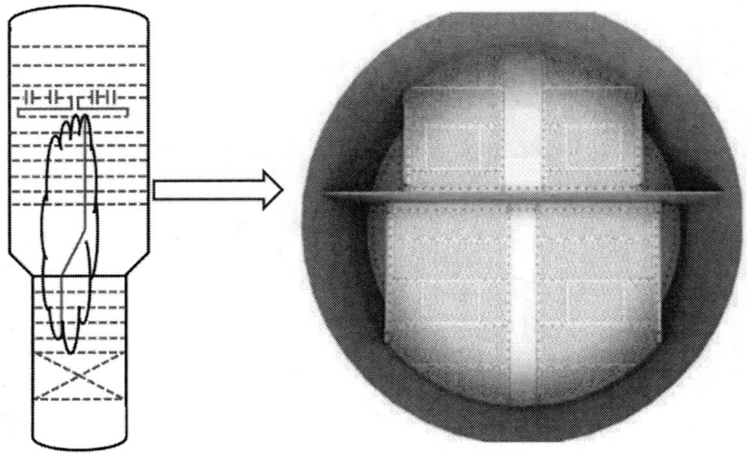

GTC Internal Liquid Split Distributer

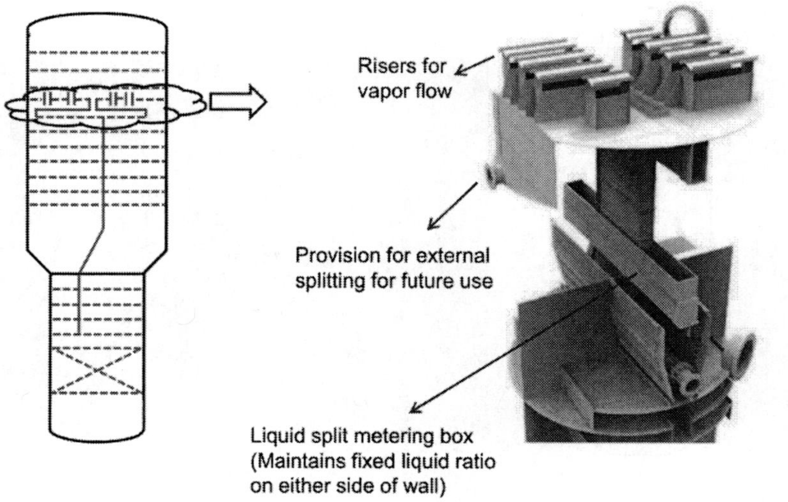

Case Study 2 : Grassroots Mixed Xylenes Recovery Unit at TonenGeneral (TG), Japan

- TG had an existing unit which produced C7+ product for gasoline blending.
- TG wanted to separate high purity petrochemicals (Toluene, Mixed Xylenes) from the feed.
- TG decided for a DWC solution against a two column configuration because of lower CAPEX and lack of plot space.

Case Study 2: Mixed Xylenes Recovery

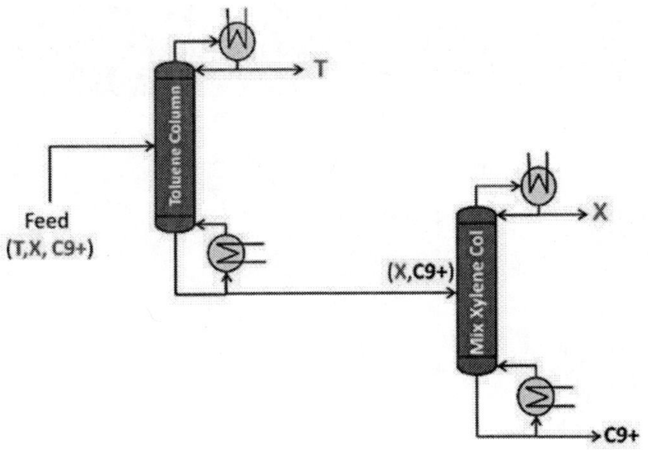

2017 AIChE® SPRING MEETING & 13TH GLOBAL CONGRESS ON PROCESS SAFETY

17 AIChE Spring Meeting, San Antonio, TX — 13TH GLOBAL CONGRESS ON PROCESS SAFETY — GTC Technology - Engineered to Innovate® — TonerGeneral

GT-DWC℠ for Mixed Xylenes Recovery

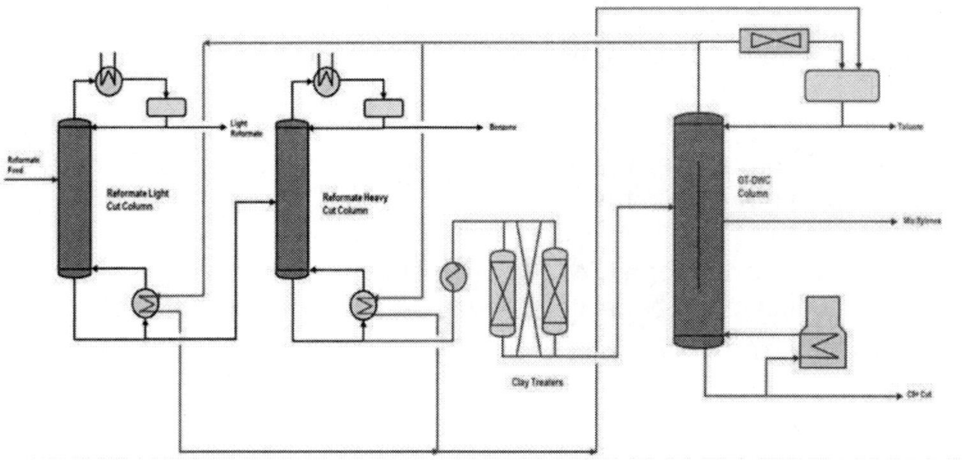

17 AIChE Spring Meeting, San Antonio, TX — 13TH GLOBAL CONGRESS ON PROCESS SAFETY — GTC Technology - Engineered to Innovate® — TonerGeneral

Comparison: Two Column System vs DWC

Parameter	Units	Two Column Configuration	GT-DWC℠ Configuration
Mix-xylenes product	Kg/hr	29,332	29,334
C8 aromatics	wt%.	99.2	99.3
Reboiler duties	MMkcal/hr.	21.6	17.2
Operating cost savings	%	-	20.0
Capital cost	US $ MM	26.0	21.0
Steam savings due to heat integration	TPH	-	20.0

Flexibility in Design

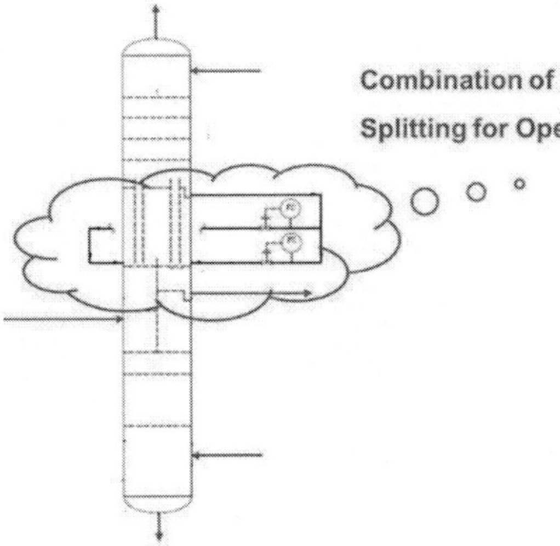

Combination of Internal and External Splitting for Operational Flexibility

2017 AIChE® SPRING MEETING & 13TH GLOBAL CONGRESS ON PROCESS SAFETY

GT-DWC℠ Column Internals at TonenGeneral

2017 AIChE® SPRING MEETING & 13TH GLOBAL CONGRESS ON PROCESS SAFETY

TonenGeneral Mixed Xylenes Recovery Unit

Installation of GT-DWC℠ Column at Chiba Refinery, Japan

Column Operational since March 2016

GT-DWC℠ Internals Manufacturing - Korea

GT-DWC℠ Trays

Testing of GT-DWC℠ Internal Liquid Split Distributer

2017 AIChE® SPRING MEETING & 13TH GLOBAL CONGRESS ON PROCESS SAFETY

17AIChE Spring Meeting, San Antonio, TX — 13TH GLOBAL CONGRESS ON PROCESS SAFETY — GTC Technology Engineered to Innovate® — TonenGeneral

Heat Integrated GT-TDWC℠

- Conventional side draw column provides limited opportunity for heat integration
- Top dividing wall column gains the thermodynamic efficiency, plus affords meaningful heat integration.

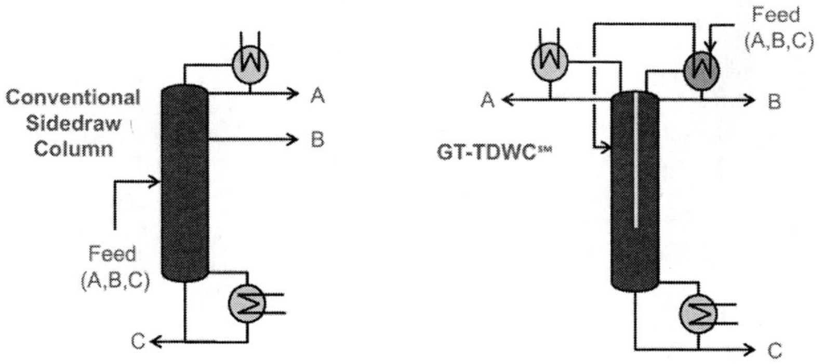

17AIChE Spring Meeting, San Antonio, TX — 13TH GLOBAL CONGRESS ON PROCESS SAFETY — GTC Technology Engineered to Innovate® — TonenGeneral

Case Study 3: GT-TDWC℠ for Food Grade Hexane Production (FGH) for BPCL, India

FGH production requires a narrow cut of an intermediate product.

Options for application

1. Conventional 2 Column System : DIH + FGH column
2. Advanced Distillation Option : GT-TDWC

2017 AIChE® SPRING MEETING & 13TH GLOBAL CONGRESS ON PROCESS SAFETY

17AIChE Spring Meeting, San Antonio, TX | 13TH GLOBAL CONGRESS ON PROCESS SAFETY | GTC Technology · Engineered to Innovate® | TonerGeneral

Conventional Option: DIH + New FGH Column

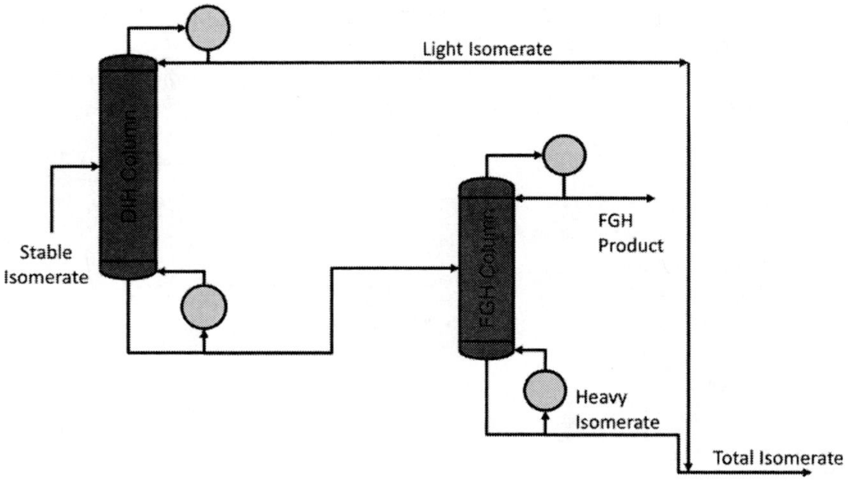

2017 AIChE® SPRING MEETING & 13TH GLOBAL CONGRESS ON PROCESS SAFETY

17AIChE Spring Meeting, San Antonio, TX | 13TH GLOBAL CONGRESS ON PROCESS SAFETY | GTC Technology · Engineered to Innovate® | TonerGeneral

GT-TDWC℠ Column for Food Grade Hexane Production

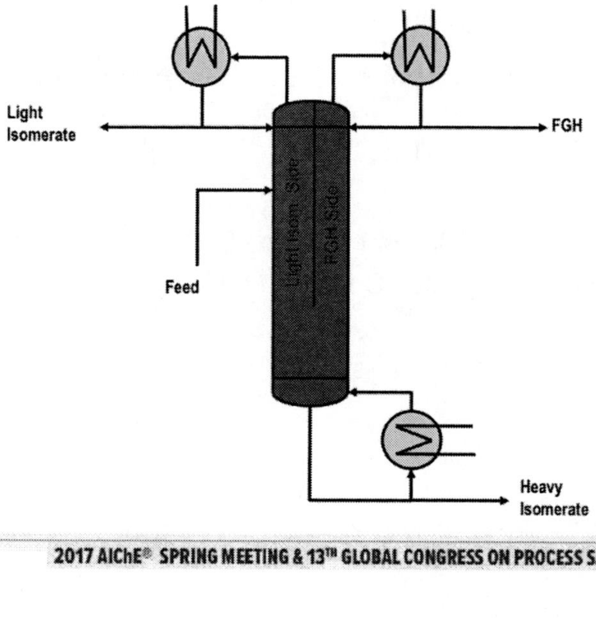

2017 AIChE® SPRING MEETING & 13TH GLOBAL CONGRESS ON PROCESS SAFETY

Comparison : GT-TDWC vs Conventional 2 Column System

	Conventional 2 Column System		GT-TDWC
	DIH Column	New FGH Column	
MP Steam Consumption (Tons/Hr)	26.20	8.50	26.46
	Total = 34.7		
Operating Cost (MM USD/year) Steam only	9.69		7.36
Savings (MM USD/year) Steam only	-		2.34

Top Dividing Wall Column Installation at BPCL, India

Case Study 4: GT-LPG Max℠ for LPG Recovery

Revamp of Existing LPG Recovery Unit in Texas to GT-LPG Max

Client Objectives:

- Propane recovery in LPG > 97%
- No Refrigeration

Case Study 4: GT-LPG Max℠ for LPG Recovery

Product Class	Product Specifications	Feed Composition	
		Component	Wt%
Fuel Gas 3.8%	Minimize Propane	Hydrogen	0.0%
		Methane	0.3%
		Ethylene	0.0%
		Ethane	3.0%
		H_2S	0.5%
		CO_2	0.0%
LPG Product 11.8%	Propane > 90 wt% Ethane < 1 wt% Vapor Pressure < 208 psig @ 100°F	Propene	0.1%
		Propane	11.7%
I-C4 Product 8.3%	Propane < 3.0 wt% N-C4 < 4.0 wt%	i-Butane	8.2%
		i-Butene	0.1%
		1-Butene	0.1%
N-C4 Product 19.6%	Iso-Pentane < 3.0 wt% Iso-Butane < 5.3 wt%	n-Butane	19.6%
		Tr2-Butene	0.0%
Debutanizer Bottoms 56.5%		Heavies	56.6%

Lights are 3.7% of Total Feed

Considerable C5 and Heavy Components

2017 AIChE® SPRING MEETING & 13TH GLOBAL CONGRESS ON PROCESS SAFETY

17AICHE Spring Meeting, San Antonio, TX — 13TH GLOBAL CONGRESS ON PROCESS SAFETY — GTC Technology Engineered to Innovate — TonenGeneral

Conventional Design of LPG Recovery Unit

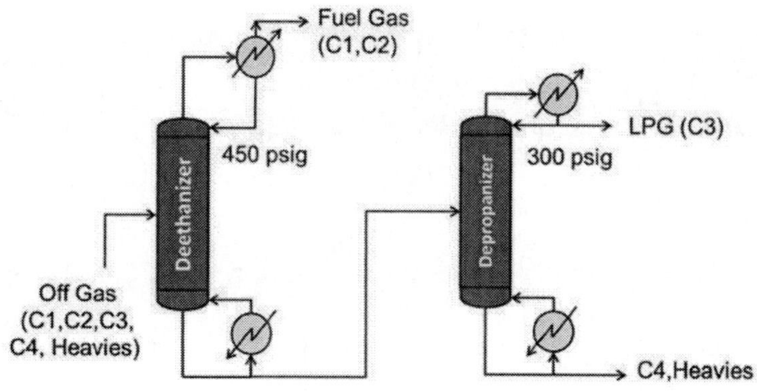

- LPG Recovery – 55%
- Energy – 22.0 MMBtu/hr

17AICHE Spring Meeting, San Antonio, TX — 13TH GLOBAL CONGRESS ON PROCESS SAFETY — GTC Technology Engineered to Innovate — TonenGeneral

GT-LPG MAX℠ - Combined Absorption & Distillation in a Single Column

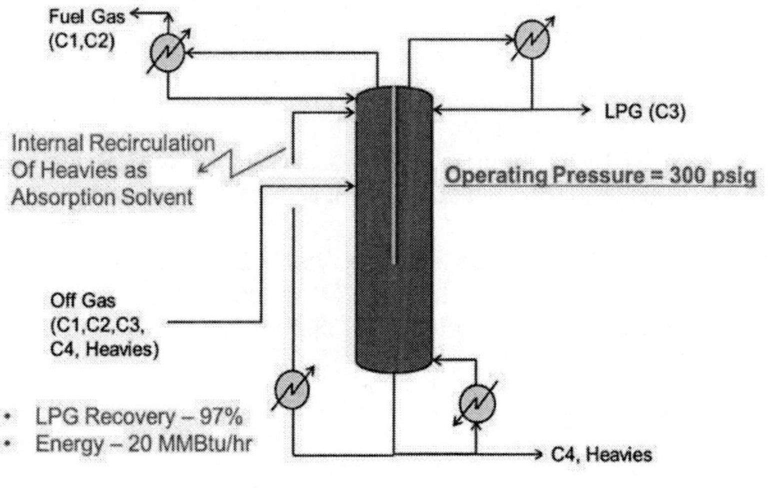

- LPG Recovery – 97%
- Energy – 20 MMBtu/hr

2017 AIChE® SPRING MEETING & 13TH GLOBAL CONGRESS ON PROCESS SAFETY

Project Economics of GT-LPG MAX℠

Variables	Existing Configuration	GT-LPG Max℠
Overall Propane Recovery %	55 %	97 %
Total Duty MM kcal/hr	22.0 MMBtu/hr	20.0 MMBtu/hr

• Net Benefit/Year	$MM	10.1
• Total Installed Cost	$MM	15.0
• Simple Pay Back	Months	18.0

Summary

- Unconventional distillation applications are an overlooked means to reduce refinery energy consumption
  - GT-DWC reduces 20–30% OPEX through energy savings
  - GT-DWC reduces 20–30% CAPEX by requiring a single column for multi-component separation
- Advanced Distillation schemes can offer CAPEX and OPEX reduction of more than 40% by combining multiple unit operations and heat integration opportunities.

Improve Energy Efficiency of DWCs With Overhead Heat Integration

Manish Bhargava
Licensing Manager
GTC Technology

Norihito Suzuki
Staff Engineer
TonenGeneral Sekiyu K.K.

479915 Distillation of High Purity Products from Feeds Containing Multiple Azeotropes

Abstract:

The basic design principles as well as startup and operation of a distillation systems to be utilized for separating multiple azeotropes for high value materials. We will discuss the determination of critical design factors as well as selection of operating conditions. Details design regarding fabrication, assembly, startup, and operation will be provided.

Cases involving known as well as variable feed compositions will be discussed. The importance of flexible operating conditions will be examined and the relationship to system and operating costs.

Component Trapping with VLE Uncertainty: Principles and Troubleshooting

By

Paul M. Mathias,[1] Henry Z. Kister,[1] Bruce Parker,[2] Lydia Narvaez,[3] Thomas Schafer,[4] Alan Erickson[4]

[1] Fluor Corporation, 3 Polaris Way, Aliso Viejo, CA 92698

[2] Eli Lilly and Company, Lilly Corporate Center, Indianapolis, IN 46221

[3] Fluor Daniel Caribbean, Inc., Suite 500 St. 2 No. 14 Metro Office Park Guaynabo, Puerto Rico 00969

[4] Koch Modular Process Systems, LLC, 45 Eisenhower Drive, Suite 350, Paramus, NJ 07652

Presented at the Kister Distillation Symposium, AIChE Spring Meeting
San Antonio, Texas, March, 2017

Component Trapping with VLE Uncertainty:

Principles and Troubleshooting

Paul M. Mathias,[1] Henry Z. Kister,[1] Bruce Parker,[2] Lydia Narvaez,[3] Thomas Schafer,[4] Alan Erickson[4]

[1] Fluor Corporation, 3 Polaris Way, Aliso Viejo, CA 92698

[2] Eli Lilly and Company, Lilly Corporate Center, Indianapolis, IN 46221

[3] Fluor Daniel Caribbean, Inc., Suite 500 St. 2 No. 14 Metro Office Park Guaynabo, Puerto Rico 00969

[4] Koch Modular Process Systems, LLC, 45 Eisenhower Drive, Suite 350, Paramus, NJ 07652

Summary

Intermediate components tend to accumulate near the middle of a distillation column in many chemical and petroleum separations. Their accumulation may lead to off-spec products, corrosion, plugging, or periodic cycling. The most common method of removing these intermediate components is by taking them out in one or more a side draws. The addition of such side draws may not go far enough to achieve the desired product specs. If the side-draw approach fails, the expensive addition of a new tower may be required.

The question of whether a side draw is sufficient to remove the accumulating intermediate components, or whether an additional column is needed, depends on the relative volatilities of the intermediate components at tower conditions. Often, there is a high degree of uncertainty in the correlation of the volatilities of these components. The Margules Uncertainty Analysis method is valuable in analyzing these situations, and in guiding engineers to the most effective path.

The case analyzed here is an acetonitrile-water separation tower with t-crotonaldehyde and propionitrile as the intermediate components. There are uncertainties in the correlation of the volatilities of these intermediate components in this system. In the light of these uncertainties, our analysis addresses the question of whether the tight product specs can be achieved in one tower with a side draw or whether more than one tower is needed. The analysis shows that depending on the product specs and the tolerance for a high reflux ratio, there would be situations where one can be certain that one tower will work, while in other cases more than one tower is needed to assure achievement of the required product specs. In the present case, a single distillation tower is judged to be too risky, and hence the more conservative 3-column solution was chosen.

Introduction

This paper studies the options to purify an aqueous solution of ≈ 12 wt% acetonitrile (ACN) by distillation, and to recover about 90% of the ACN at the azeotropic composition (about 83 wt% at 1 atmosphere or 101.3 kPa). The separation is difficult due to intermediate components [propionitrile (PPN) and crotonaldehyde (T-CROTON] that build-up in the distillation column. The solution options are a single column with a side draw, or a more complex flowsheet with multiple columns. An example of a multiple-column solution is to remove all of the water from the ACN, PPN and T-CROTON via a pressure swing distillation, and then separate the dry ACN from the PPN and T-CROTON. The multiple-column solution is the safe solution, but costs more than the single column with a side draw in terms of both capital cost and operating costs. Here we quantify the risk of the single-column solution by introducing uncertainty analysis into the process modeling.

The problem of intermediate components in distillation is a general problem of component trapping that has been discussed extensively by Kister.[1,2] In some cases, the top temperature is too cold and the bottom temperature is too high to allow these components to leave the column at the same rate as they enter. Having nowhere to go, these components accumulate in the column, which causes problems like off-spec products, flooding, cycling slugging, and corrosion. Cycling on a variable time period (typically with frequencies between two hours and two weeks) is a common symptom indicating unsteady state component trapping. Diagnoses of the problem include studying time variation of temperature profiles, taking internal samples, shooting gamma scans, and simulation studies. Solutions comprise raising or lowering the temperature (usually at the expense of recovery), taking side draws, or removing the problem components from the feed. Diagnoses and solutions have been discussed in detail by Kister.[1,2] Here we apply uncertainty analysis to the problem caused by trapping of intermediate components in ACN purification.

Modeling of the distillation separations requires gaining quantitative understanding of the uncertainties in the vapor-liquid equilibrium (VLE) and applying these uncertainties to the problem at hand. While many technologists have highlighted the need to recognize the impact of uncertainties in property models on process design and plant operability[3,4,5,6,7] and several solutions have been suggested and evaluated in the literature,[8,9,10,11,12,13,14] uncertainty analysis is hardly used in chemical-engineering design practice. In fact, Kim et al.[15] observed, "Its practical implementation in a variety of scientific and engineering fields has typically seen less emphasis than it deserves." Mathias[16] recognized that liquid-phase nonideality is the major source of phase-equilibrium uncertainty, and developed an intuitive and easy-to-apply uncertainty-analysis method based upon treating the mixture, for the purpose of activity-coefficient perturbation, as a set of pseudobinaries described by the Margules equation. The Margules perturbation method has been applied to several case studies, including propylene-propane superfractionator, where small changes in relative volatility and product purity have a large effect on design;[16] a dehexanizer column that removes pentane and hexane from a heavier slate of components that are mostly aromatic and the system contains many components (≈ 60);[16] the "textbook problem" of acetone/benzene/chloroform separation, which revealed the relationship between uncertainty analysis and residue-curve analysis;[17] the butanol/water separation, which demonstrated the extension to liquid-liquid equilibrium (LLE) and is a case where the thermodynamic model is inadequate;[18] and ethyl acetate purification, which showed the importance of fitting multicomponent data for complex systems.[19]

The Margules uncertainty methodology used here follows that of the previous studies.[16,17,18,19] The first step is detailed data regression to obtain the best model fit in the region of application (atmospheric pressure, about 101.3 kPa), and quantitative estimation of the correlation uncertainties. This is a necessary first step. The data regression here is mainly applied to the water/ACN binary mixture, with PPN and T-CROTON at low concentration. In the second step, activity coefficients are varied within the process model to relate the design uncertainties to the confidence limits of the correlated phase equilibria.

Margules-Based Activity-Coefficient Perturbation Scheme

The perturbation scheme for the mixture nonideality has been published by Mathias,[16] and hence is only briefly presented here. The approach uses the approximation that the perturbed activity coefficients can be guided by the Margules equation.

$$\ln(\gamma_i) = \ln(\gamma_i^m) + \ln(\gamma_i^p) \tag{1}$$

where γ_i is the activity coefficient of component i after perturbation, γ_i^m is the activity coefficient calculated by the chosen model (the "best" model in the region of interest should be used), and γ_i^p represents the perturbation obtained through the estimated model uncertainty. It should be emphasized that γ_i^m is calculated from the "best" activity-coefficient model, and is not in any way constrained by the pseudo-binary approximation. γ_i^p is given by,

$$\ln(\gamma_i^p) = \delta_i (1-x_i)^2 \left[\frac{\left|\ln(\gamma_i^m)\right|}{(1-x_i)^2 + \left|\ln(\gamma_i^m)\right|} \right] \tag{2}$$

As previously noted,[16] Equation 2 is phenomenological, but has desired characteristics. The value of $\ln(\gamma_i^p)$ is small when γ_i^m is close to unity (i.e., low departures from ideal behavior), and also is low when the mole fraction of component *i* is close to unity. For components at concentrations close to infinite dilution and with activity coefficients significantly different from unity, the fractional change in γ_i resulting from the perturbation asymptotically approaches $\{exp(\delta_i) - 1\}$. In order to apply the perturbation method, the value of δ_i for each component must be related to the estimated uncertainty in its activity coefficient. This thorough data analysis is a necessary element of the perturbation scheme and is discussed in detail later in the paper.

Correlation of VLE in the Water/ACN/PPN/T-CROTON System

As in the previous uncertainty-analysis case studies,[16,17,18,19] the perturbation scheme has been implemented as a user activity-coefficient model in the NRTL-RK property option in Aspen Plus V8.8. The NRTL-RK property option uses the NRTL activity-coefficient model[20] and the Redlich-Kwong equation of state[21] for the vapor phase. The pure-component properties come from the Aspen Plus V8.8 pure-component database. This study is at a pressure of 1 atm (101.3 kPa), the temperatures of interest are approximately 70-100 °C, and at these temperatures the correlations for the vapor pressures of these four compounds from the Aspen Plus database are expected to be accurate to better than 1 %. The reason for this confidence is that the Aspen Plus pure-component properties come from the DIPPR 801 database,[22] and the correlated vapor pressures have been reported to be accurate to <1%. The Redlich-Kwong equation estimates the vapor-phase fugacity coefficient, but this thermodynamic quantity is expected to be close to unity and accurately predicted since the pressure of interest is low, ≈1 atm. For this application, the phase-equilibrium uncertainty is largely determined by the uncertainties in the component activity coefficients.

The NRTL model has two temperature-dependent binary parameters (α_{ij}^{N} and τ_{ij}). α_{ij}^{N} is a symmetric binary parameter and τ_{ij} is an asymmetric binary parameter,

$$\tau_{ij} = a_{ij} + b_{ij}/T + e_{ij}\ln(T) + f_{ij}T \tag{3}$$

$$\alpha_{ij}^{N} = c_{ij} + d_{ij}(T - 273.15) \tag{4}$$

where, T is the temperature in Kelvin.

The K-value of component i in a mixture (K_i) at vapor-liquid equilibrium (VLE) is defined to be the ratio of the vapor and liquid mole fractions at equilibrium, and the relative volatility between components i and j (α_{ij}) is the ratio of the K-values.

$$K_i \equiv \frac{x_i}{y_i} \tag{5}$$

$$\alpha_{ij} \equiv \frac{K_i}{K_j} \tag{6}$$

Note that the superscript "N" has been added to the NRTL parameter α_{ij}^{N} in order to distinguish it from the relative volatility, α_{ij}.

The Data Regression System (DRS) in Aspen Plus V8.8 has been used to obtain optimum values of the NRTL binary parameters at pressures at about 1 atm (101.3 kPa), which corresponds to the temperature range of ≈ 70-100°C. Phase-equilibrium data were obtained from the NIST-TDE[23] database, and key VLE data have been measured as part of this effort. The water/ACN binary is well-studied systems since a large number of datasets are available, however, the four binaries consisting of PPN and T-CROTON with water and ACN have only a few data. In fact, data for systems with PPN are very poorly covered, and hence this study has mainly focused on the effect of T-CROTON as the representative intermediate component.

Table 1 presents the 14 components considered in the Aspen Plus simulation together with their boiling points. The table also shows the estimated infinite-dilution activity coefficients in water and ACN at a representative temperature of 75°C; these correlations/estimations are discussed later in this paper. The last column of Table 1 designates components other than water and ACN as "light" or "heavy." The "light" components (1-4) have normal boiling points lower than that of ACN. These components will always have K-values greater than unity, and hence will split predominately to the distillate in the distillation column. The "heavy" components (10-14) have normal boiling points greater than that of water, and also are fairly ideal with water (infinite-dilution activity coefficients close to unity). These "heavy" components will always have K-values less than unity, and therefore will largely split to the bottoms stream of the distillation column. The "intermediate" components have normal boiling points higher than that of ACN, and form fairly ideal mixtures with ACN, consequently their K-values are likely to be below unity at the top of the column (in the vicinity of the water/ACN azeotrope). Further, the "intermediate" components have T_b values close to that of water and also water infinite-dilution activity coefficients significantly greater than unity (see Table 1), and therefore their K-values will be greater than unity at the bottom of the column. As a result, the intermediate components have K-values that cross unity and tend to build-up in the column.

Table 1. Components in the acetonitrile purification Aspen Plus model.

Component Name	T_b (°C)	Component ID	γ^∞ at 75°C in		Impurity Category
			Water	ACN	
1. Acetaldehyde	21.0	ACETALD	9.0	1.1	Light
2. Dichloromethane	39.8	DCM	235.9	1.2	Light
3. n-Propionaldehyde	49.0	PROPALD	32.2	1.3	Light
4. Acrolein	52.7	ACROLEIN	17.1	1.1	Light
5. Acetonitrile	81.6	ACN	11.7	1.0	-
6. cis-Crotonaldehyde	88.7	C-CROTON	32.5	1.4	Intermediate
7. Propionitrile	97.4	PPN	41.6	1.0	Intermediate
8. Water	100.0	WATER	1.0	6.1	-
9. trans-Crotonaldehyde	102.5	T-CROTON	32.5	1.4	Intermediate
10. Acetic Acid	117.9	ACETACID	1.3	1.9	Heavy
11. 2-Methoxyethanol	124.4	MEOXYETH	2.7	2.0	Heavy
12. Morpholine	128.0	MORPHOL	1.1	1.0	Heavy
13. Acetamide	221.2	ACETAMID	1.0	1.0	Heavy
14. 4-Ethylmorpholine	239.4	ACETMOR	1.1	1.0	Heavy

Crotonaldehyde has two isomers, *cis* and *trans*. The study of McGreer and Page[24] found that cis-crotonaldehyde is far less chemically favored than trans-crotonaldehyde, by about 1:50. Hence, although cis-crotonaldehyde is available in the simulation file, its properties are not considered reliable, and crotonaldehyde has been treated as entirely the trans isomer (T-CROTON)

It is clear that the system of interest to this study consists of 4 components: water, ACN, PPN and T-CROTON. For the other binary mixtures, the NRTL parameters were obtained by regressing data from NIST-TDE,[23] or were estimated using the UNIFAC method.[25] It is noted that the K-values of the other components (1-4 and 10-14 in Table 1) have negligible effect on this study.

Table 2. NRTL parameters for the water/ACN/PPN/T-CROTON system. See Eqs. 3 and 4 for a description of the parameters. Parameters e_{12}, e_{21}, f_{12}, f_{21} and d_{12} are always equal to zero.

Component 1	ACN	PPN	WATER	ACN	ACN	PPN
Component 2	WATER	WATER	T-CROTON	PPN	T-CROTON	T-CROTON
a_{12}	-2.34	-4.53	5.38	0	0	0
a_{21}	2.39	2.99	-2.34	0	0	0
b_{12}	1165.5	2041.1	-751.4	27.1	55.1	0
b_{21}	-199.2	-9.9	914.6	-30.0	55.1	0
c_{12}	0.45	0.41	0.3	0.3	0.3	0.3

Table 2 presents the NRTL parameters for the water/ACN/PPN/T-CROTON system. The water/ACN binary is well covered by experimental VLE data. The binaries of water with PPN and T-CROTON are also covered by VLE and LLE data; it is noted that these two binaries have regions of liquid-liquid immiscibility, but the concentrations of PPN and T-CROTON in the process are too low to exceed their solubility limit. No data were found in the literature for the other three binaries. Data for the water/ACN/T-CROTON ternary were measured as part of this project.[26] The NRTL parameters for the ACN/PPN are estimated based upon UNIFAC predictions.[25] Generally, UNIFAC predictions are not accurate, so future measurement was considered pending outcome of this study. The NRTL parameters for the PPN/T-CROTON binary have been defaulted to zero. UNIFAC estimations[25] indicate that this is a fairly ideal system (infinite-dilution activity coefficients ≤ 1.2). Minor components have a negligible effect on overall mixture nonideality, and it was decided that the assumption of ideality is the preferred choice.

Key comparisons of model results to experimental data are provided next in order to estimate the VLE uncertainties. Figure 1 compares the calculated relative volatility in the water/ACN binary to data from 9 sources available in NIST-TDE.[23] The model describes the data, including the azeotropic region, accurately, and there is no indication of model bias. The dashed lines show $\pm$ 20% from the model calculation. It is noted that that the water/ACN binary is not a significant cause of design uncertainty, and no uncertainty perturbation has been applied to the water/ACN binary. Figure 2 presents comparisons between model and data[26] for the K-value of T-CROTON at low concentration (effectively infinite dilution) in the water/ACN binary mixture at 1 atm (101.3 kPa). The dashed lines indicate $\pm$ 30% from the model results. Most of the data points fall within the dashed lines, and this may be considered to be the combined uncertainty (about 3 standard deviations) of the model K-values for T-CROTON. The vertical line in Figure 2 at 83 wt% shows the ACN/water azeotrope, which corresponds to the distillate composition from the column, and it is clear that both the model and experimental data indicate that the K-value of T-CROTON drops below unity before reaching the top of the column, which, of course, is the reason that T-CROTON is an intermediate component. Figure 3, which shows the K-value of PPN, is analogous to Figure 2. While the calculated results are equivalent for PPN and T-CROTON, there are no experimental data available for PPN, and hence the perturbation studies here are limited to T-CROTON, with the expectation that they will be approximately applicable to PPN.

Figure 4 provides a visual indication of the effect of $\delta_{T\text{-}CROTON}$ (hereafter simply referred to as δ, since the Margules perturbation in this work is only applied to T-CROTON). The bottoms concentration is nearly-pure water and the distillate concentration corresponds to the water/ACN azeotrope. Referring to Eq. 2 and Figure 2 it is clear that for a given value of δ, the perturbation at the bottom of the column (higher T-CROTON activity coefficient) will be higher than at the top of the column, and that a value of $\delta = 0.4$ is a "full" perturbation, corresponding effectively to the 30% uncertainty in Figure 2. Note that only positive values of δ are relevant to the study because increased values of δ will tend to increase the concentration of T-CROTON in the distillate, which makes it more difficult to meet the design goal of low T-CROTON concentration in the distillate.

Two design studies have been conducted to evaluate the effect of T-CROTON VLE uncertainty on the performance of the single column, and these are described next.

Design Study 1 – Single Column with Side Draw

Figure 5 presents the single column with side draw used for the first set of design studies. The feed flow is 770 kg/hr and is an aqueous mixture containing 12 wt% ACN, and with minor impurities. The impurity levels of PPN and T-CROTON in the feed are 100 ppmw and 20 ppmw, respectively. The column has 35 theoretical stages including a reboiler and total condenser. The distillate-to-feed ratio is set at 0.13, which corresponds to $\approx$ 91% ACN recovery at the azeotropic composition. The reflux ratio is varied to obtain a T-CROTON concentration in the distillate of 1 ppmw. The feed stage is varied from 12 to 24. The base case is the feed to stage 20, and for this case the calculated reflux ratio is 7.0. Figure 6 shows the vapor and liquid concentration profiles (weight fraction) for ACN, PPN and T-CROTON in the distillation column for the base case. The concentration of ACN rises with stages toward the condenser (here in decreasing stage number) and quickly reaches the water/ACN azeotropic concentration just a few stages above the feed stage. The concentration profiles of PPN and T-CROTON show the characteristics of intermediate components since they both have concentration peaks below the feed stage. Note that the concentration peak of T-CROTON is about three orders of magnitude above the distillate and bottoms concentrations. The distillate and bottoms concentrations of both PPN and T-CROTON are both small since most of these two components are removed in the side draw. Figure 7 emphasizes these points by showing the K-value profile of T-CROTON alongside its concentration profile. Note that the relative values of the vapor and liquid concentrations switch as the K-value moves from less than one at the top of the column (lower stage numbers) to greater than one at the bottom of the column.

Figure 8 studies how the reflux ratio needs to be adjusted to attain the spec of 1 ppmw T-CROTON in the distillate as the perturbation parameter (δ) and the number of rectification stages are varied. Increasing the number of rectification stages is helpful in reducing the required reflux ratio, and this can inferred from Figure 6 (and seen in Figure 8) because the concentrations of PPN and T-CROTON decrease monotonically in the rectification section of the column, but the effect is relatively weak. On the other hand, increasing δ (increased activity coefficient and therefore K-value of T-CROTON) results in a strong increase in the required reflux ratio. Very large reflux ratios cause problems, for example requiring large-diameter columns and significant increases in reboiler duty, and are usually not accepted in industrial process-design practice

Figure 9 studies the effect of the draw rate for the base case (Figure 5 with reflux ratio of 7.0 and feed to stage 20). It is clear that varying the draw rate has a strong effect on the concentration of T-CROTON in the bottoms (because it changes the total amount of T-CROTON removed in the side draw), but it has only a weak effect on the concentration of T-CROTON in the distillate. Since the main goal here is to reduce the concentration of PPN and T-CROTON in the distillate, the next study is done without the side draw.

Design Study 2 – Single Column with no Side Draw

Figure 10 presents the single column with no side draw used for the second set of design studies, and with the total number of stages increased from 35 to 40. Other design conditions are the same as in Figure 5. Figure 11 is qualitatively the same as Figure 8, which supports the conclusion that the side draw affects the bottoms concentration of T-CROTON, but has only a weak effect on the key result we desire, which is minimizing the concentration of T-CROTON in the distillate.

Figure 12 presents the results of a study where the reflux ratio has been fixed at 10.0, and the perturbation parameter (δ) and the number of rectification stages are varied. Similar to Figure 8, Figure 12 shows that it is beneficial to increase the number of rectification stages as this reduces the concentration of T-CROTON in the distillate. However, the key result is that the concentration of T-CROTON in the distillate is a very strong function of the uncertainty in the VLE. As δ increases not only does the concentration of T-CROTON in the distillate increase beyond the target value (1 ppmw), but it can also exceed the feed concentration (20 ppmw), in effect concentrating undesired impurities in the product!

The single-column solution was rejected as being too risky.

Conclusions

The purpose of this study was to apply the Margules uncertainty method to an industrially important component-trapping case, i.e., purification of acetonitrile used as a solvent in the pharmaceutical industry. Small amounts of T-CROTON and PPN turn a simple distillation into a very challenging problem. The study first focused on the VLE correlation, and determined that the key uncertainty is the K-values of the intermediate components, PPN and T-CROTON. Since K-value data are only available for T-CROTON, this component was chosen as the representative intermediate key component. The first set of studies indicated that the target of 1 ppmw T-CROTON in the distillate can be met, at least in the simulation model, but at a significant increase in reflux ratio if the T-Croton K-value is increased within the established uncertainty. The second set of studies concluded that increasing the K-value of T-CROTON within the established uncertainty caused a significant increase in the T-CROTON concentration in the distillate, and the product concentration could even exceed the feed concentration. The three-column solution was the eventual low-risk process solution selected.

List of Tables

List of Figures

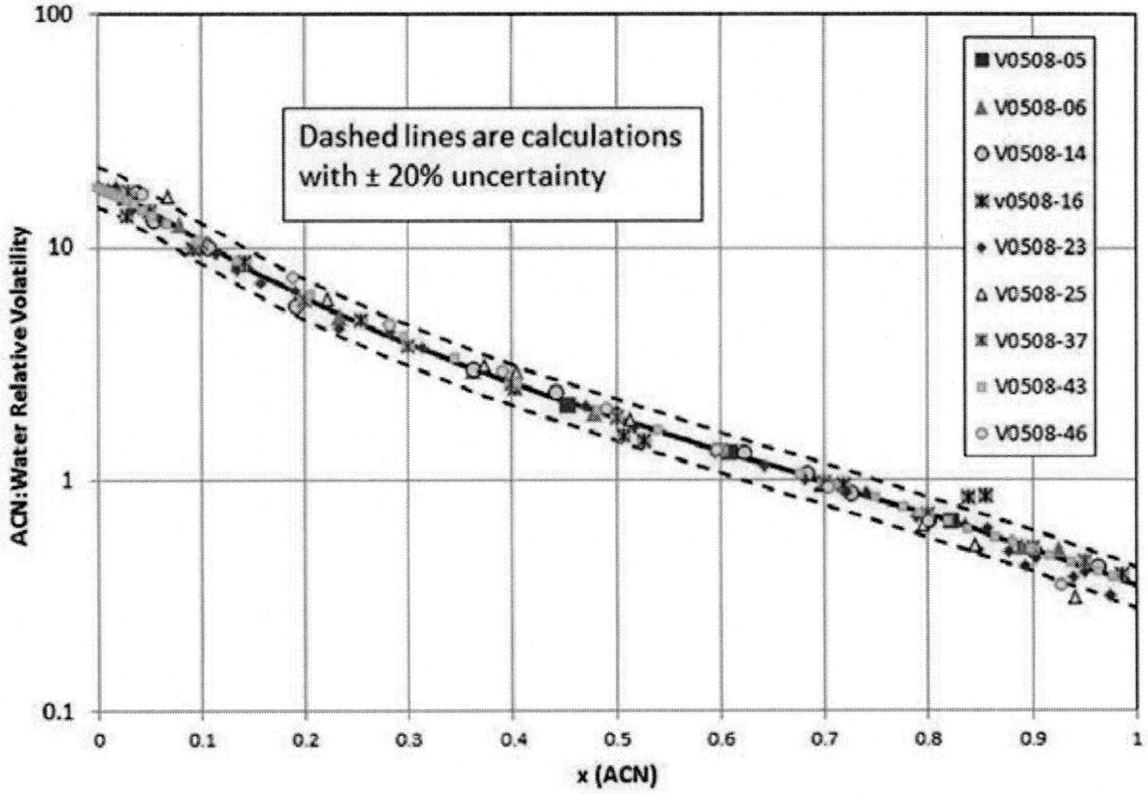

Figure 1. Water/ACN relative volatilities at 1 atm (101.3 kPa). Comparison of model calculations to 9 data sources from NIST-TDE.[23]

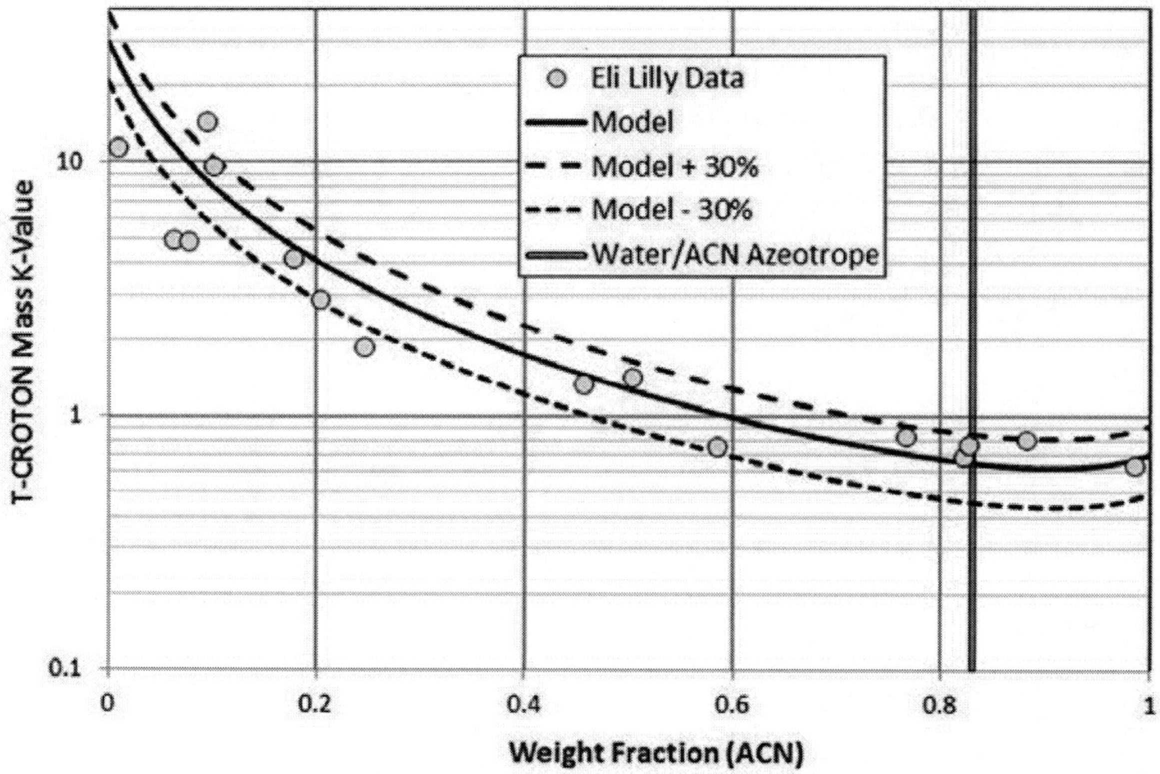

Figure 2. K-Value of T-CROTON at low concentration in water/ACN at 1 atm (101.3 kPa). Comparison of model calculations to Eli Lilly data.[26] The dashed lines show ± 30% from the model results. The vertical line indicates the water/ACN azeotrope.

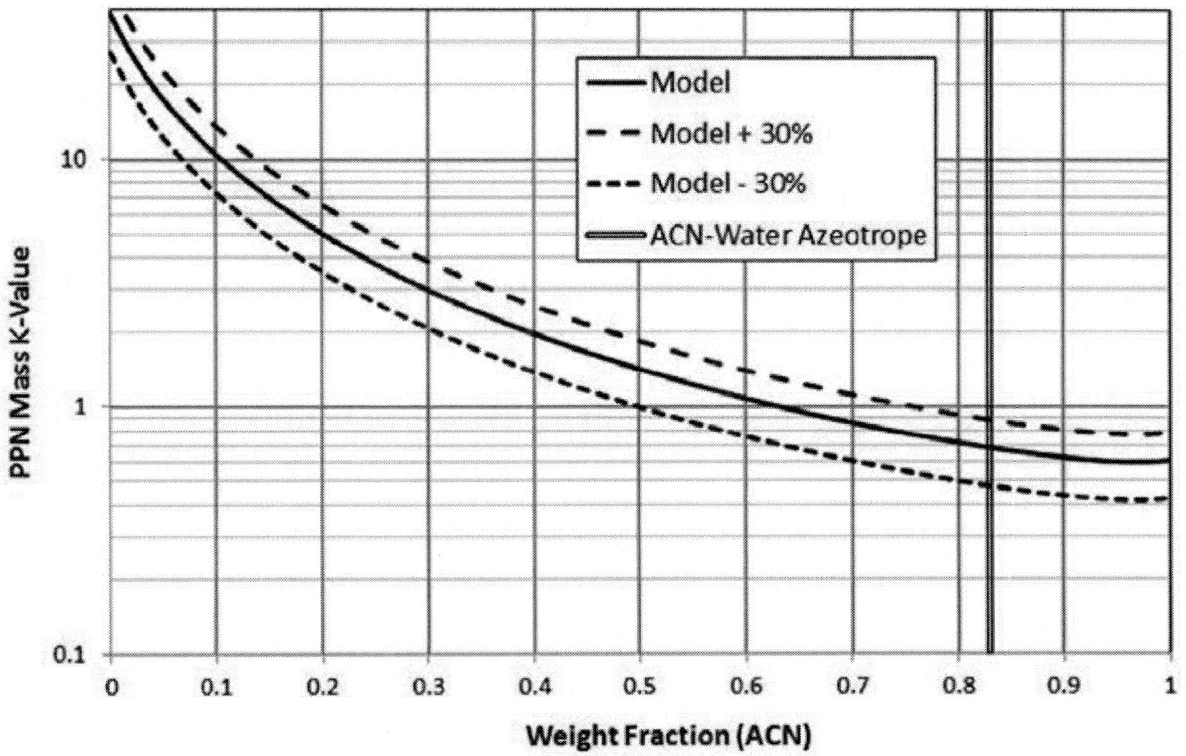

Figure 3. K-Value of PPN at low concentration in water/ACN at 1 atm (101.3 kPa). No experimental data are available. The dashed lines show ± 30% from the model results. The vertical line indicates the water/ACN azeotrope.

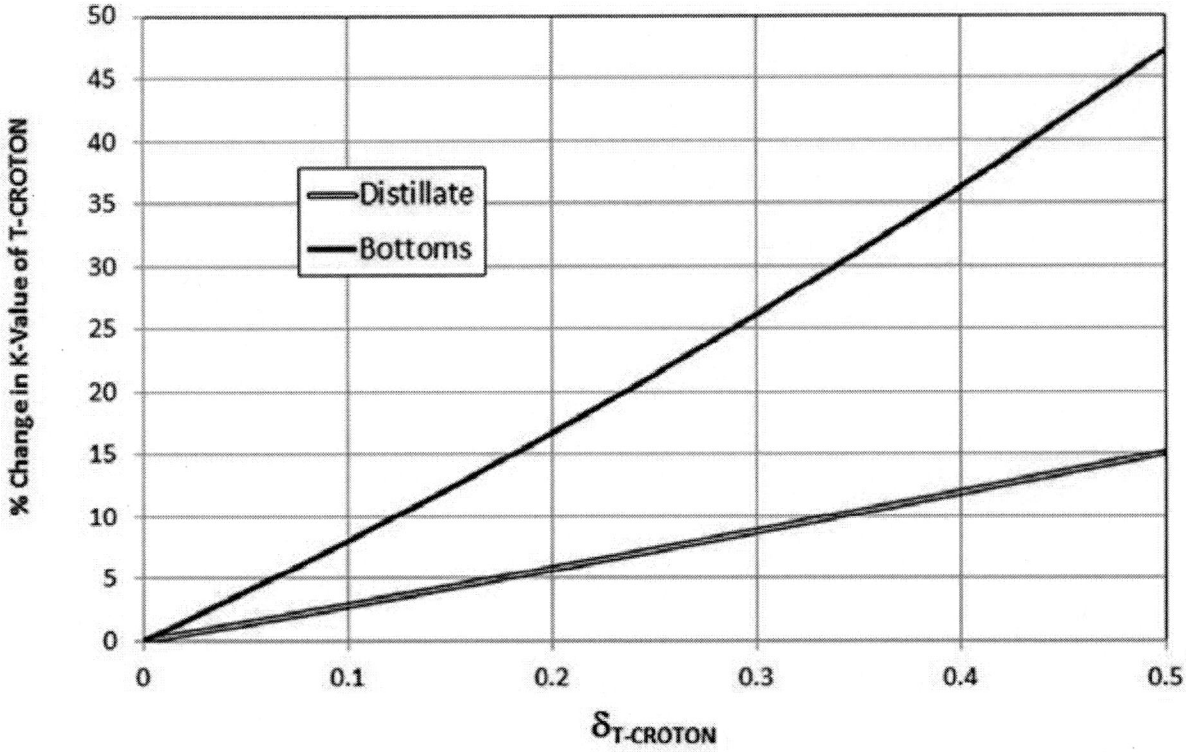

Figure 4. Effect of T-CROTON perturbation parameter (δ) on the percentage change in the K-value of T-CROTON. The bottoms concentration is nearly-pure water and the distillate concentration corresponds to the water/ACN azeotrope.

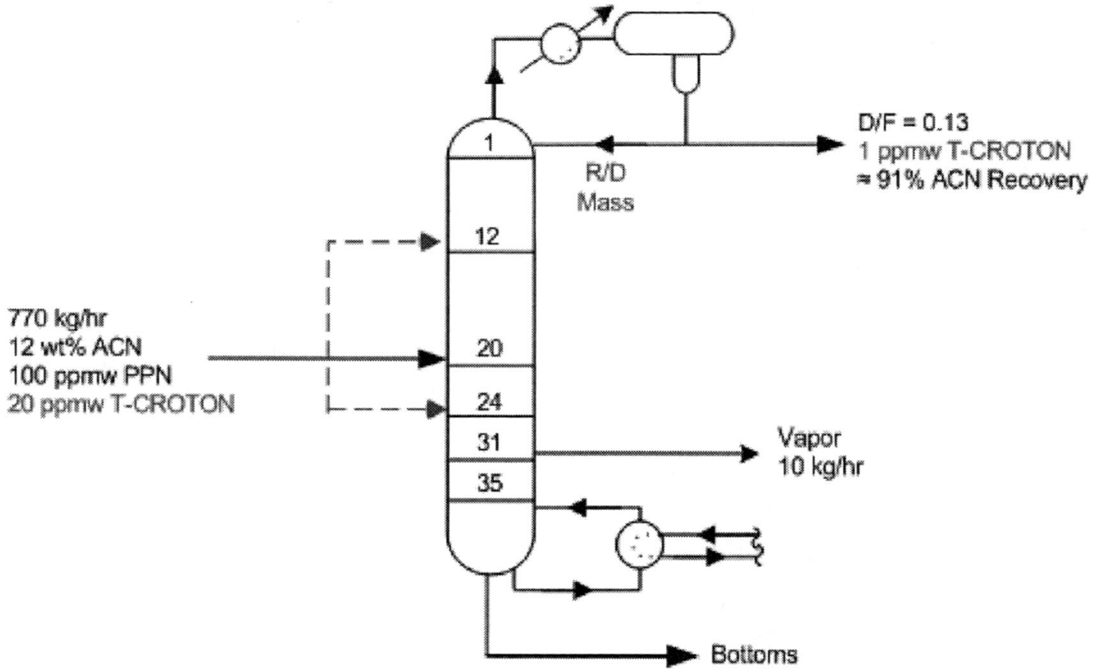

Figure 5. Single column with side draw used for perturbation studies-1. The column operates at 1 atm (101.3 kPa).

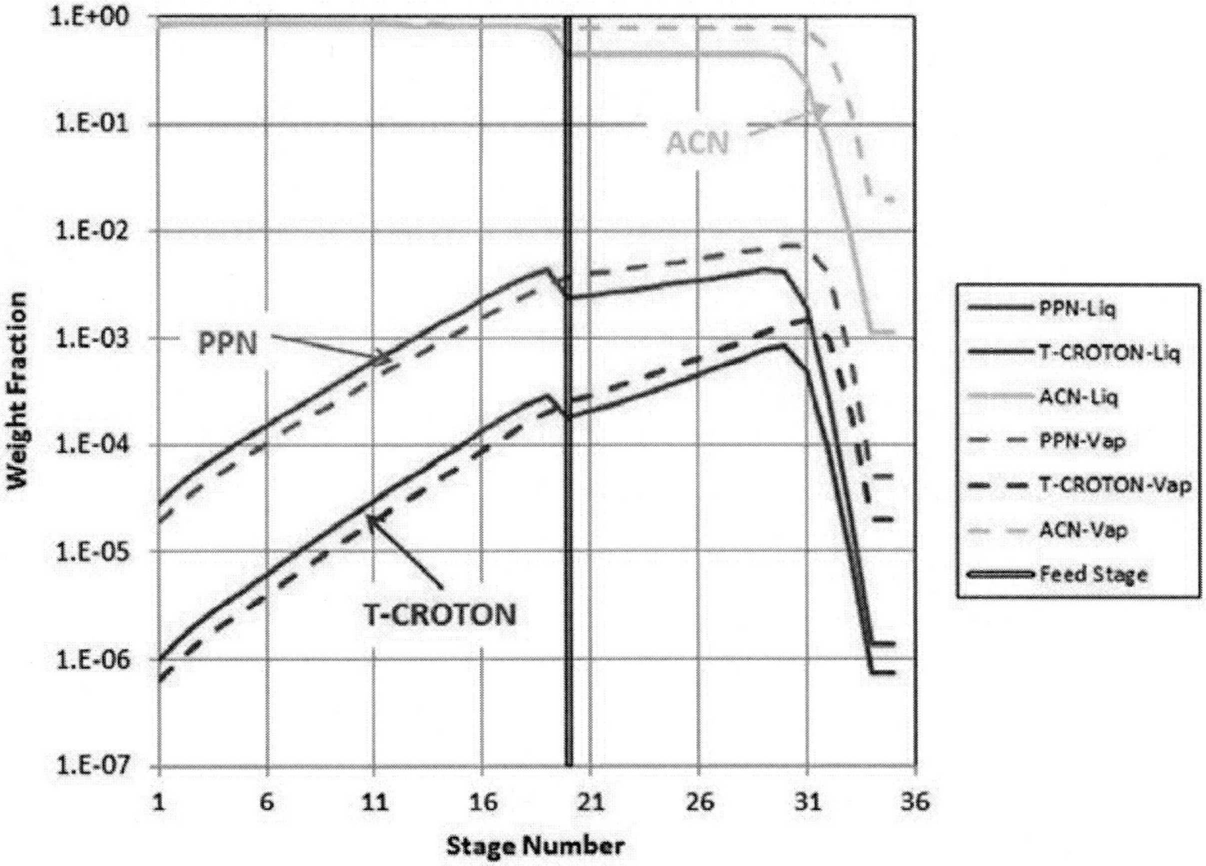

Figure 6. Vapor and liquid concentration profiles for ACN, PPN and T-CROTON in distillation column corresponding to Figure 5, and with feed to stage 20.

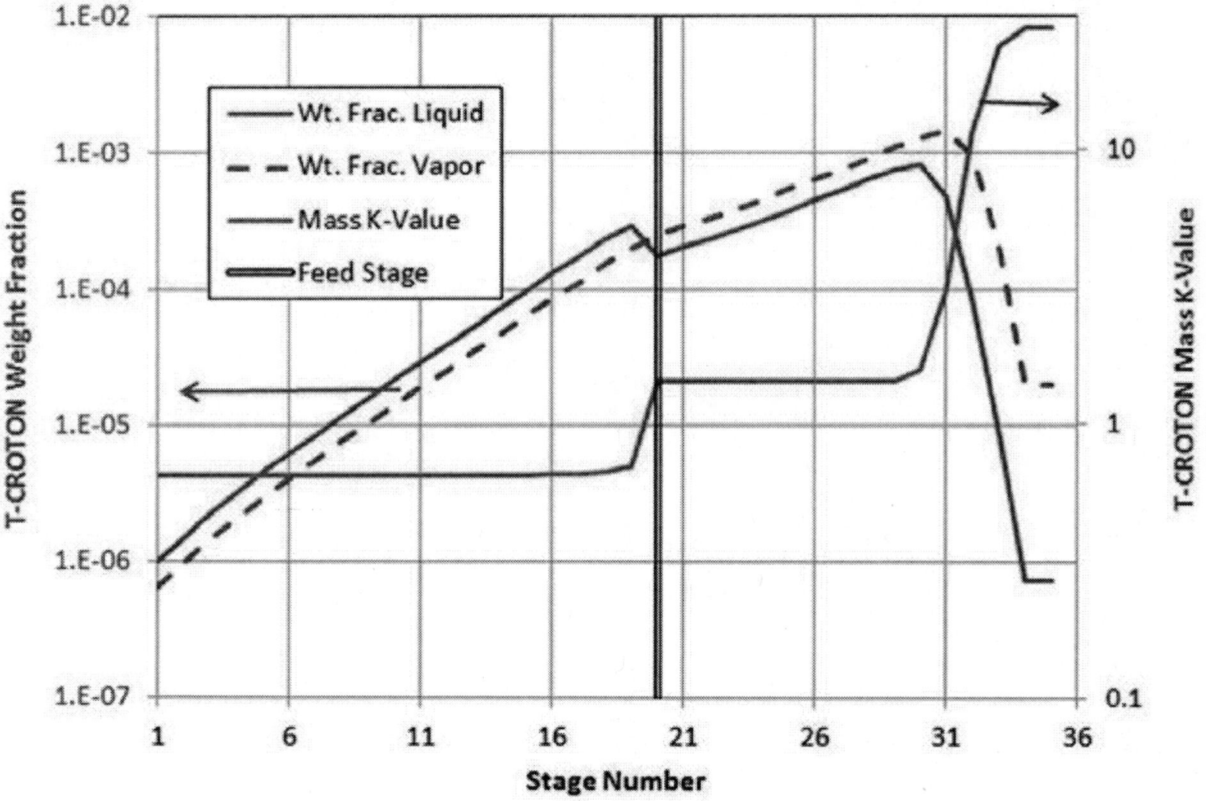

Figure 7. Mass-fraction and K-value profiles of T-CROTON in the base case. The specifications are the same as those in Figure 6.

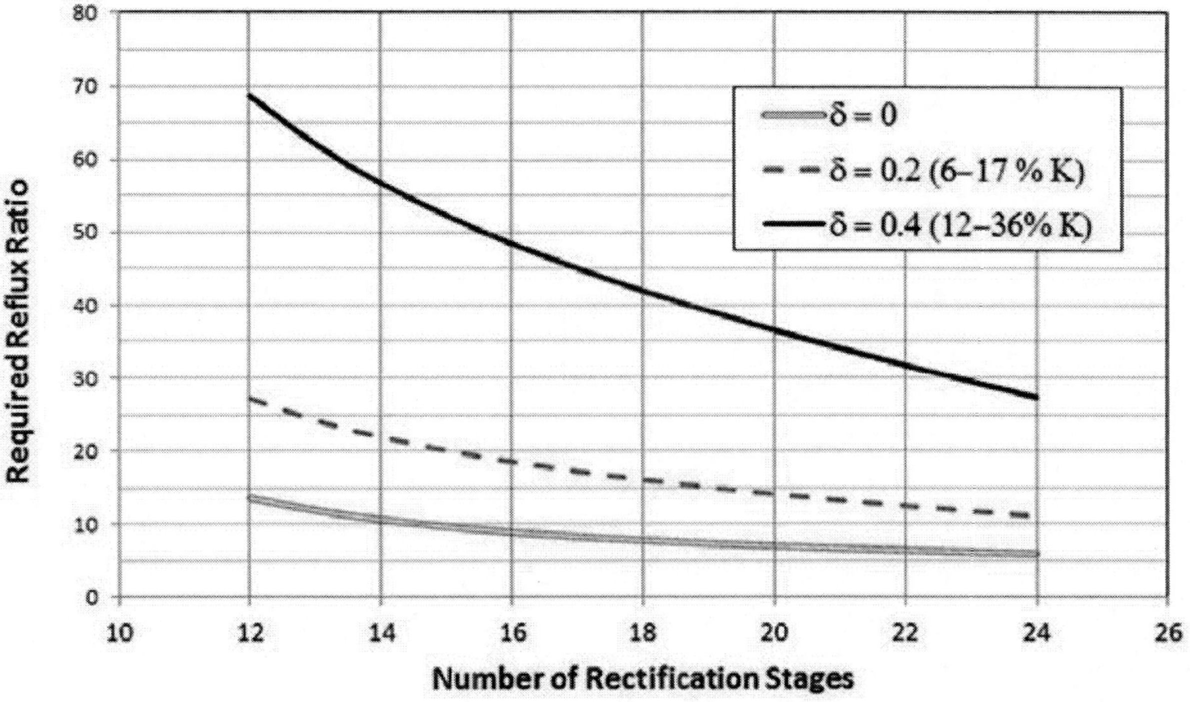

Figure 8. Effect of perturbation parameter (δ) and number of rectification stages on the reflux ratio required to meet the spec of 1 ppmw T-CROTON in the distillate. See Figure 5 for details.

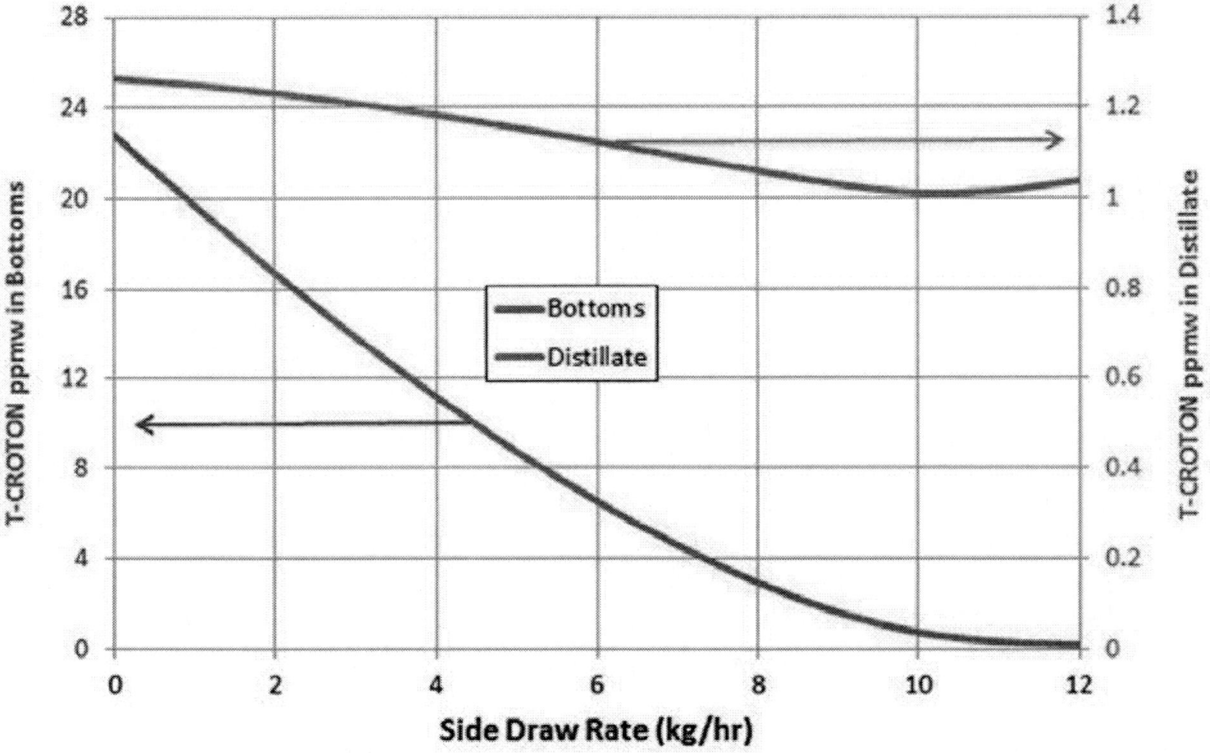

Figure 9. Effect of modifying side-draw rate. The feed is at stage 20 and the reflux ratio is fixed at 7.0.

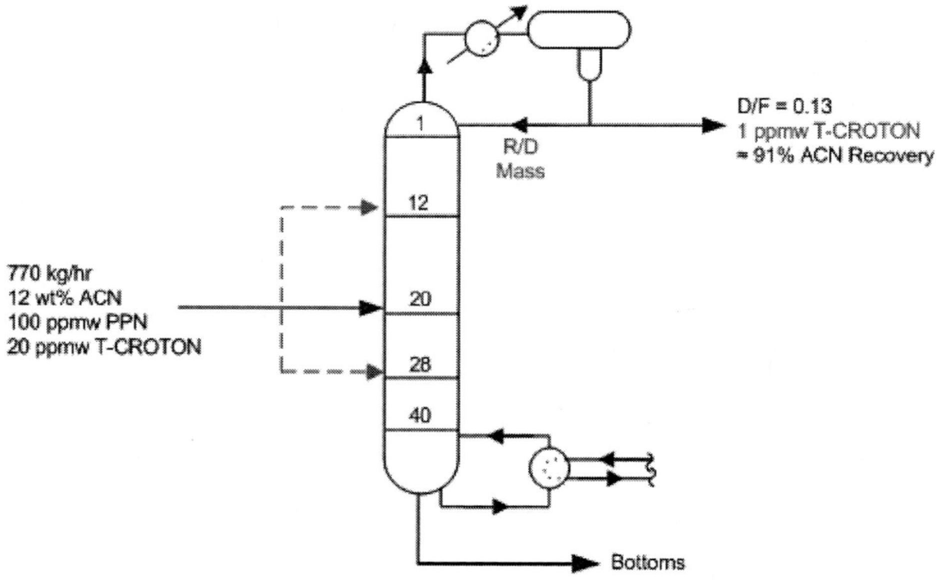

Figure 10. Single column with perturbation studies-2. The difference from Figure 5 is that the side draw has been eliminated.

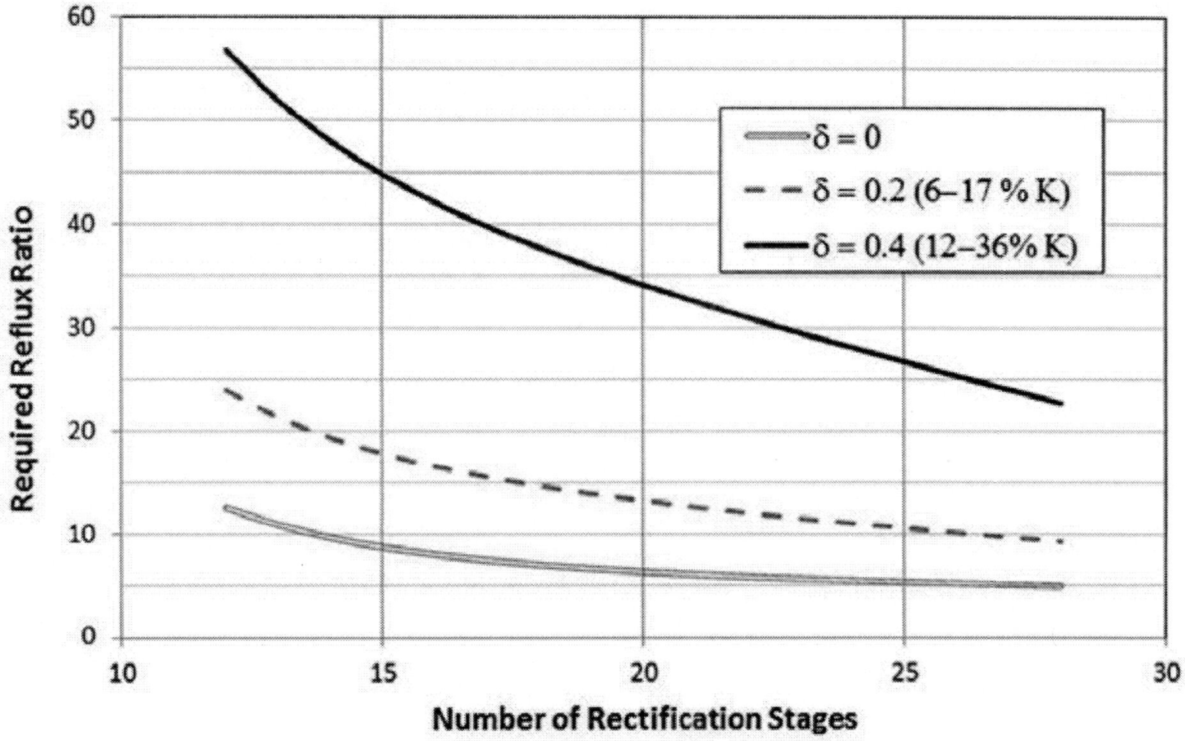

Figure 11. Repeat of study shown in Figure 8 with no side draw and total number of stages increased from 35 to 40 (See Figure 10).

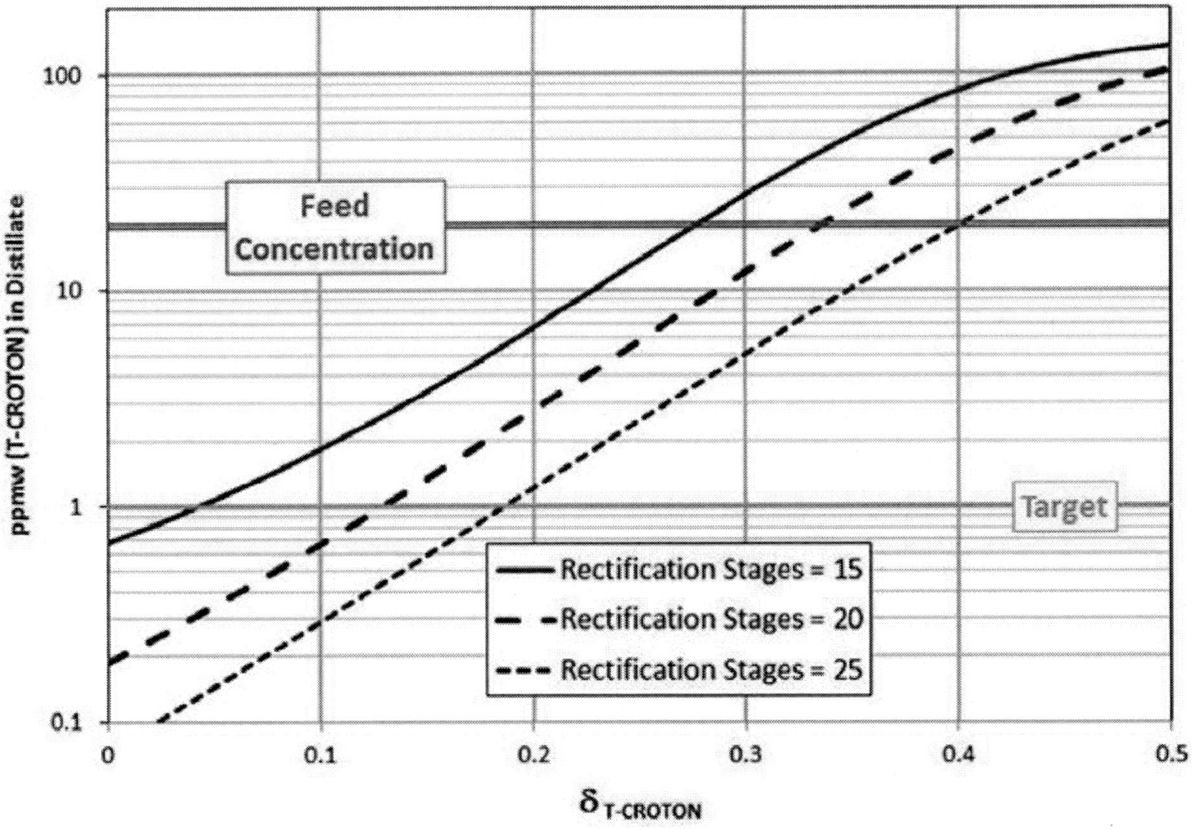

Figure 12. Effect of perturbation parameter (δ) and number of rectification stages on concentration of T-CROTON in distillate. Referring to Figure 10, the reflux ratio is fixed at 10.

Literature Cited

[1] Kister, H. Z. Components Trapping in Distillation Towers: Causes, Symptoms and Cures," *Chem. Eng. Prog*. **2004**, 100, 22-33.

[2] Kister, H. Z. *Distillation Troubleshooting*. John Wiley and Sons, Inc., Hoboken, NJ, **2006**.

[3] Streich, M.; Kistenmacher, H. Property Inaccuracies Influence Low Temperature Designs, *Hydrocarb. Proc*., **1979**, 58, 237-241.

[4] Mah, R. S. H. Effects of Thermophysical Property Estimation on Process Design, *Computers Chem. Eng*., **1977**, 1, 183-189.

[5] Zudkevitch, D. Imprecise Data Impacts Plant Design and Operation, *Hydrocarbon Process*. **1975**, 54, 97–103.

[6] Zudkevitch, D.; Gray, R. D. Impact of Fluid Properties on the Design of Equipment for Handling LNG, *Adv. Cryog. Eng*. **1975**, 20, 103–123.

[7] Larsen, A. H. Data Quality for Process Design, *Fluid Phase Equilib*., **1986**, 29, 47–58.

[8] Zeck, S. Thermodynamics in Process Development in the Chemical Industry - Importance, Benefits, Current State and Future Development, *Fluid Phase Equilib*., **1991**, 70, 125–140

[9] Macchietto, S.; Maduabeuke, G,; Szcepanski, R. Exact Determination of Process Sensitivity to Physical Properties, Fluid Phase Equilib., **1986**, 29, 59-67.

[10] Reed, M. E.; Whiting, W. B., Sensitivity and Uncertainty of Process Designs to Thermodynamic Model Parameters: A Monte Carlo Approach, *Chem. Eng. Commun*. **1993**, 124, 39–48.

[11] Whiting, W. B. Effects of Uncertainties in Thermodynamic Data and Models on Process Calculations, J. *Chem. Eng. Data*, **1996**, 41, 935–941.

[12] Whiting, W. B.; Vasquez, V. R.; Meerschaert, M. M. Techniques for Assessing the Effects of Uncertainties in Thermodynamic Models and Data, *Fluid Phase Equilib*. **1999**, 158–160, 627–641.

[13] Xin, Y.; Whiting, W. B. Case Studies of Computer-Aided Design Sensitivity to Thermodynamic Data and Models, *Ind. Eng. Chem. Res*. **2000**, 39, 2998–3006.

[14] Whiting, W. B.; Vasquez, V. R.; Meerschaert, M. M. Techniques for Assessing the Effects of Uncertainties in Thermodynamic Model and Data, *AIChE J*., **1997**, 43, 440–447.

[15] Kim, S. H.; Kang, J. W.; Kroenlein, K.; Magee, J. W.; Diky, V.; Frenkel, M. Online Resources in Chemical Engineering Education: Impact of the Uncertainty Concept from Thermophysical Properties, *Chem. Eng. Educat*., **2013**, 47, 48-57.

[16] Mathias, P. M., Sensitivity of Process Design to Phase Equilibrium – A New Perturbation Method Based Upon the Margules Equation. *J. Chem. Eng. Data*, **2014**, 59, 1006-1015.

[17] Mathias, P. M. Effect of VLE Uncertainties on the Design of Separation Sequences by Distillation – Study of the Benzene-Chloroform-Acetone System. *Fluid Phase Equilib*., **2016**, 408, 265-272.

[18] Mathias, P. M. Effect of Phase-Equilibrium Uncertainties on the Separation of Heterogeneous Azeotropes – Application to the Water/1-Butanol System. *J. Chem. Eng. Data*, **2016**, 61, 4077-4084.

[19] Mathias, P. M.; Kister, H. Z. Effect of Phase-Equilibrium Uncertainties on Ethyl Acetate Purification, submitted for publication to *J. Chem. Eng. Data*, February **2017**.

[20] Renon, H.; Prausnitz, J. M. Local Compositions in Thermodynamic Excess Functions for Liquid Mixtures. *AIChE J.*, **1968**, 14, 135-144.

[21] Redlich, O.; Kwong, J. N. S. On the Thermodynamics of Solutions. V An Equation of State. Fugacities of Gaseous Solutions. *Chem. Rev.*, **1949**, 44, 233-244.

[22] Rowley, R. L.; Wilding, W. V.; Oscarson, J. L.; Yang, Y.; Zundel, N. A. DIPPR® *Data Compilation of Pure Chemical Properties, Design Institute for Physical Properties*, AIChE, New York, NY, **2010**.

[23] Frenkel, M.; Chirico, R. D.; Diky, V.; Yan, X.; Dong, Q.; Muzny, C. ThermoData Engine (TDE): Software Implementation of the Dynamic Data Evaluation Concept. *J. Chem. Inf. Model.*, **2005**, 45, 816-838.

[24] McGreer, D.E.; Page, B. D. cis-Crotonaldehyde and Related Compounds. *Can. J. Chem.*, **1969**, 47, 866-867.

[25] UNIFAC is an estimation method based upon group contributions. Several UNIFAC models are available in Aspen Plus V8.8. In the present study, since mainly approximate estimations are required, the base UNIFAC model has been used. Fredenslund, Aa.; Gmehling, J.; and Rasmussen, P. *Vapor-Liquid Equilibria Using UNIFAC, A group-Contribution Method*, 3rd Impression, Elsevier, Amsterdam, **1977**.

[26] Parker, B. Data measured in Eli Lilly Corporate Center Laboratory, August, **2016**.

The Successful Application of a Partition Flow Spray Tray (PFST®) in C4 Extractive Distillation

Zhuxiang Wang, Enxia Shang, Wei Hong, Tianjin ChuangJu Technology Co. Ltd, Tianjin, P. R. China

Yan Zhuang, Advanced Column Technology, Perth, Australia

The crude C4 hydrocarbons production, as a by-product of the ethylene industry, exceeds 6 million tonnes per annum in China. As the first step in crude C4 upgrade, Butadiene is recovered from the crude C4 using extractive distillation. The extractive distillation columns used to extract butadiene from the crude C4 presently have issues of high pressure drop, lower than design throughput, poor separation and high steam consumption due to its high foaming tendency of the solvent and a high liquid to vapour load ratio[1][2].

This paper presents three industrial case studies using ChuangJu's proprietary partition flow spray tray (PFST®) design to solve the above mentioned problems. The PFST® is designed with a partition plate to catch the liquid that has been sprayed out of the vertical sieve caps after mixing with the vapor. The liquid is then channeled down to the next tray by a multiple down comer and suspended settling trough design to handle the high liquid load, thus eliminates the problem of liquid back mixing on the tray. The settling troughs provide extra residence time for liquid and vapour disengagement. The spray action of the vertical sieve cap suppress foaming, thus reducing the tendency of vapour underflow back mixing and flooding caused by foaming. With this design, the vapor and liquid phases come to contact counter-currently with the maximum mass transfer driving force, everywhere on the tray with little liquid flow gradient or concentration gradient, so the tray efficiency is increased.

In the first case study, PFST® trays were used to revamp two butadiene extractive distillation columns and one de-gassing column for the Ningbo Haode Chemicals. Before the revamp, float valve trays were used and the maximum throughput achieved was only 70% of the design value. After the revamp, the feed rate is running at 140% of the design value with stable operation.

The other two C4 separation plants, which were designed, fabricated, installed and commissioned by ChuangJu, are running at above design capacities with good separation and recovery performance.

Keywords: partition flow spray tray, C4 extractive distillation, butadiene, foaming system, PFST®

INTRODUCTION

ChuangJu's R&D on tray column design focuses on solving the problems of high pressure drop, vapour phase uneven distribution caused by the hydraulic gradient on the tray, liquid back mixing, as well as flooding caused by foaming etc. Within the series ChuangJu spray tray designs, the PFST® tray is especially suitable for the C4 extractive distillation in that it can minimise liquid phase back-mixing under a high liquid load and it can also supress foam formation[3]

Figure 1: A picture of a PFST® tray

THE DEVELOPMENT OF THE PARTITION FLOW SPRAY TRAY

Conventional trays, such as a F1 float valve tray, can be struggling to handle a system with a high liquid load, susceptible to foaming and components with very close relative volatilities. The reasons are:-

- The liquid and vapour are contacted in a cross-flow pattern. Due to the concentration gradient of the liquid flow, the mass transfer driven force is reduced along the liquid flow path, which leads to a lower tray efficiency.
- High liquid load leads to higher hydraulic gradient which can cause uneven vapour flow distribution.
- For a foaming susceptible system, the conventional tray design does not have a foaming suppression mechanism, other than adding foaming suppression agents, so it is easy to run into a flooding.
- Some of the conventional tray designs have "mass transfer dead zones" near the column wall where there exist stagnant flow, side flow and circular flow, which can lower the tray efficiency.

The PFST® tray design uses multiple down comers and spray contactors to handle a high liquid load and foaming issue. By using a partition plate to separate the portion of liquid that has been contacted with the vapor, from the liquid coming directly from the tray above, thus reduces liquid phase back mixing, which increases mass and energy transfer efficiency.

THE MAIN CHARACTERISTICS OF THE HIGH PERFORMANCE PFST® TRAY

Multiple Down Comer Design

For a high liquid load situation, multiple down comer design can shorten the length of liquid flow path, thus reduce the hydraulic gradient on the tray, which make the vapor rise more evenly across the tray. It also can effectively control the liquid layer height on the tray within an optimized range.

Spray Type Contactors

The spray type contactor offers higher mass and energy transfer efficiency by dispersing the liquid phase into a large amount of small droplets, which creates more interfacial contact areas.

Foaming Suppression

The spray type tray has a built-in foaming suppression mechanism which makes it a preferable choice for a foaming susceptible system. The droplets sprayed out of the spray contactors falling back to the tray floor which can break some of the bubbles.

Reduced Liquid Back Mixing

By using a partition plate to catch the liquid that spray out of the spray contactors, ie to separate the portion of liquid that has been in contact with the vapor, from the portion coming directly from the tray above, there is much less chance for liquid phase back mixing. There is also much reduced hydraulic gradient, concentration gradient and temperature gradient on the tray. As a result, each of the vertical spray contactors has the same point efficiency achieved by a uniform and high mass transfer driving force across the tray.

Lower Pressure Drop

In a spray type tray, the vapor flow does not have to pass through the liquid layer on top of the tray. The vapor phase pressure drop only need to counteract the kinetic energy used to lift the portion of liquid into the spray contactors, which leads to lower pressure drop compared with a conventional tray operating in the froth operating regime. In a low vapor load situation, the pressure drop of the spray tray is comparable to that of a F1 float valve tray. In a high vapor load situation the pressure drop of the spray tray is 20 to 30% lower than that in a F1 float valve tray. The higher the vapor load, the higher the pressure drop reduction by using a spray tray.

Higher Capacity

In a spray type tray, because of the structure of the spray contactor, the liquid-vapor mixture sprays out of the cap either horizontally or downwardly, as per the figure depict. This increases the traveling path and time for the disengagement of the phases, thus reducing liquid entrainment. With the same tray open areas, the allowable vapor velocity can be 1.5 to 2.0 time higher than that in a conventional tray, so the capacity can be much higher.

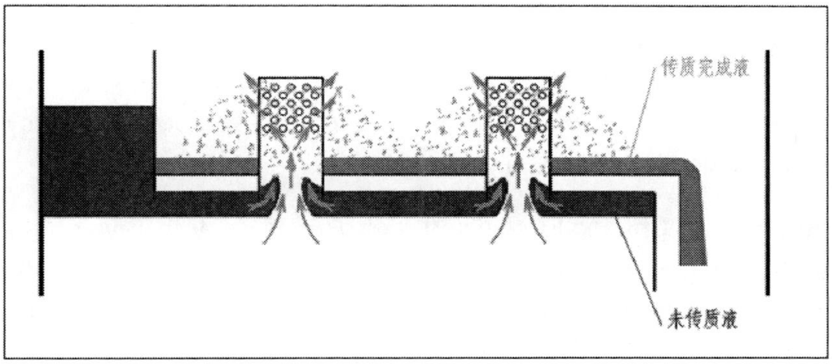

Figure 2: How does a Partition Flow Spray Tray Work

CASE STUDY 1: NINGBO HAODE CHEMICAL INDUSTRY LTD.

附图六 宁波昊德 C4 分离单元各塔

The Ningbo Haode owns and operates a facility that produce 47,000t/a of isobutene. The facility consists of a C4 separation, n-butene isomerisation unit, a MTBE synthesizing unit and a MTBE cracking to isobutene unit. The capacity of the C4 separation unit is 120,000t/a, using an extraction solvent of MEK and NFM[4]. This system uses a high liquid to vapour load; is susceptible to foaming; and is difficult to fractionate due to similar relative volatilities. The extractive distillation towers A&B and the degassing tower are the key pieces of equipment, as seen in the Figure above.

The original design of the columns used a fixed valve tray. After the start up in March 2016, the unit could only run at 70% of the design capacity. ChuangJu was invited to investigate and solve the problem. After site visit with ChuangJu's technical personnel, they identified the key the issues causing the low throughput, including: the liquid flow cross section area is too small; the tray separation efficiency is low, and there is severe foaming tendency which make columns easily flooded.

Based on these findings, ChuangJu selected their PFST® tray to revamp the two extractive distillation columns and the solvent degassing column. The revamped columns were started-up successfully in June 2016. The achieved throughput is increased by 70% compared with that before the revamp, and has been running at 30% above its design capacity since then and there is still upward potential for higher throughput.

It can be seen from Table 1, both the steam consumption and the solvent loss due to entrainment are reduced. The sheducled maintenance time was alse reduced by three times a year. The facility can consistently produce on-spec product throughout since the completion of the revamp. The combined economic savings equals RMB13.7 million, while the revamp cost was only RMB 2 million, which means the pay-back time is less than two month.

Table 1: Ningbo Haode C4 Processing Column Revamping Results

Parameters		Before Revamp	After Revamp
Type of tray		Fixed Valves	PFST®
Feed rate, kg/h	Distillation Columns A/B	12,560	21,600
	De-gassing Column	124,660	226,930
Temperature℃	Distillation Columns A/B	142.8	140.3
	De-gassing Column	156.8	154.4
Steam Consumption kg/t feed	Distillation Columns A/B	520	295
	De-gassing Column	53.8	37.5
Solvent entrainment, ppm	Distillation Columns A/B	1,000	0
	De-gassing Column	1,000	0
Product Butane Content %		96%	96%
Total olefin content%		90%	90%
Scheduled maintenance times per year		4	1

CASE STUDY TWO:- 100,000 T/A RAW C4 DOWNSTREAM PROCESSING , LUOYANG PETROCHEMICAL HONGDA INDUSTRY LTD.

Figure 4: Raw C4 Process at Louyang Petrochemical Hongda Industry Ltd

In September 2013, the Louyang Petrochemical Hongda Industry Ltd started its 100,000t/a Raw C4 Downstream Processing Facility, which includes an extractive distillation, a de-gassing and a solvent regeneration column. ChuangJu designed and manufactured all the three columns. The extractive distillation column and the de-gassing column used ChuanhJu's PFST® tray. The unit reached the design capacity and met or exceeded all products specification within 65 hours of start-up with steady operation.

Table 2: Louyang Petrochemical Hongda C4 Facility Design vs Achieved Results

	Design	Achieved Results
Capacity, t/a	100,000	100,000
Extractive Distillation Column Temperature 0C	132	118.8
De-gassing Column Temperature 0C	172.9	165
Steam saving, t/d		48
Iso-butane product Content, %	≥96	98
Olefin Content, %	≥60%	66

CASE STUDY THREE:- 300,000 T/ARAW C4 DOWNSTREAM PROCESSING UNIT, HEBEI HAITE WEIYE PETROCHEMICAL LTD.

Figure 5: 300,000t/a Raw C4 Downstream Processing Facility at Hebei Haite Weiye

In October 2014, the Hebei Haite Weiye Petrochemical Ltd started its 300,000t/a Raw C4 Downstream Processing Facility, which includes an extractive distillation, a de-gassing and a solvent regeneration columns. ChuangJu supplied PFST® trays for the extractive distillation column and the de-gassing column. The unit reached 150% of the design capacity and the tray pressure drop is averaged at 550Pa. After a successful start-up, the product quality and energy consumption all met or exceeded design specification.

CONCLUSION

The PFST® high efficiency spray tray technology has been successfully commercialised in C4 separation facilities such as the design and supply of the three distillation columns for Luoyang Petrochemical Hongda, the revamping of the extractive distillation and de-gassing columns for Ningbo Haode, as well as several other C4 downstream treatment applications.

The successful industrial application proved that the PFST® tray can handle large diameter columns with difficult process condition such as high liquid to vapour ratio and foaming, which opens further possibility for similar industrial applications.

The following points can be drawn from the two case studies:

1. By using a spray contactor the mass transfer is intensified by operating in a stable and controlled spray regime, without the risk of running into jet-flooding or high liquid entrainment.
2. By separating the fresh liquid from the tray above and the liquid coming out of the spray contactors, using a partition plate, the liquid phase back mixing is eliminated, which ensures the maximum driving forces from counter-current contact of liquid and vapour is realised.
3. The multiple down comer trough design gives more time for liquid/vapour disengagement thus reduces vapour recycle. This is especially important for a foaming susceptible system.
4. The multiple down comer design also shortens the liquid flow path, thus reduces the hydraulic gradient, which leads to an even vapour flow distribution. This design also eliminates stagnant flow, short circuiting flow as well as circular flow on the tray.

References:

[1] Zhang Long. "The Principles and Technologies for Downstream Processing of the C4 and C5 Fractionates" , The Chemical Industry Press, Beijing, China, 2011

[2] Diego B，Candido D A et al. Process for separation of butadiene by pluralatage extractive distillation: US，412845. 1978.

[3] Wang Zhuxiang, et al, "A Partition Flow No Back Mixing Spray Tray", Chinese Patent：CN 102961888 A，2013-03-13.

[4] Wang Wenhua, Ren Wanzhong, The Research and Development of Extractive Distillation on the Separation of alkanes and alkenes in Raw C4 Mixture, "The Journal of Technology and Development of Chemical Industry", 1999(2):20-23.

480648 Research and Development about a New Catalytic Distillation Structured Packing

Abstract:

Research and development about a new catalytic distillation structured packing

Our team developed a new type of modular catalytic structured packing Winpak-C. Systematic studies like theoretical simulations and experimental validations were carried out to guide the industry production, which has successfully realized the industrial application.

In the process of development, packing property, pressure drop, liquid holdups, gas loading velocity were tested. Based on which, a revised irrigated pressure drop model were proposed and fitted well with the experimental results.

In order to meet the production requirement, we optimized the dry pressure drop of Winpak-C with the help of Solidworks, Gambit, Fluent and Tecplot 360. By analyzing the flow pattern near the packing junction, we proposed that reducing the height of the reactive elements to allow the gas streamlines become more gradual. The model and the optimization method were validated by the experiments. The irrigated pressure drop and other system also proved that the model is credible.

We also compared axial and radial dispersion coefficient of the reactive elements. Based on the RTD curves which obtained by the conductometer, residence time, dimensionless variance, skewness, axial and radial dispersion coefficient can be calculated which highlight the advantages of Winpak-C compare with traditional catalytic distillation structured packing.

We will introduce some industrial application cases in different process base on our new catalytic structured packing in terms of successfully applications in MTBE production, light gaso-line etherification, isomerization reactions.

479666 Tray Operating Window - Description and Applications

Abstract:

Stable and effective operation of distillation trays are characterized by vapor and liquid loads falling within a number of limitations imposed by the tray geometry. Flooding, limits capacity of trays and can be induced by vapor and/ or liquid rates. At the low operating end, weeping, dumping and/ or flow regime transitions can severely affect the ability of a tray in achieving the desired separation. Significant amount of research has been done over the years for correlating those limitations to operating conditions, physical properties of fluids involved and tray geometry. Tray designers rely strongly on those correlations for optimizing the design of a tray based on a given set of process loads and physical constraints. Those correlations can be used to plot tray operating limits as a function of liquid and vapor loads for a given tray geometry and are often referred to as tray operating windows or stability diagrams. Those diagram are extremely useful to process engineer in setting up operating conditions or making process adjustments that ensure effective fractionation in tray towers. They are also a valuable tool for a trouble shooter when assessing poor performance of distillation towers. This paper describes the different limits plotted on typical tray operating windows and their applications in process design and operation through a few practical examples.

477798 Hidden Ties! - an Explanation for O'Connell's Success

Abstract:

Hidden Ties! – An explanation for O'Connell's success

Markus Duss, Sulzer Chemtech, Winterthur - Switzerland
Ross Taylor, Clarkson University, Potsdam, NY - USA

O'Connell's method has been widely used to estimate the performance of distillation columns equipped with trays and can be expressed as follows:

$$\eta_{Section} = 0.503 \cdot (\mu_L \cdot \alpha)^{-0.226}$$

A few attempts have been undertaken attempting to explain O'Connell's success. However, no prior explanation is entirely convincing. This presentation will show that O'Connell's simple, empirically derived correlation has ties to a more fundamental mass transfer based approach. A traditional approach determining the tray efficiency from the number of transfer units, point efficiency, tray efficiency and section efficiency is adapted and it is shown that there is less than 10% deviation from the O'Connell correlation when:

a) the number of transfer units are equal for the gas and liquid phase
b) the number of transfer units depends only on the liquid viscosity

A simple modification of the alpha-term is proposed:

$$\eta_{Section} = 0.503 \cdot \mu_L^{-0.226} \cdot \alpha^{-0.08}$$

The proposed modification leads to almost perfect agreement between a theoretical model and this modified O'Connell model when the above assumptions hold true and when $\alpha < 5$.

O'Connell's correlation should now be seen in a different light.

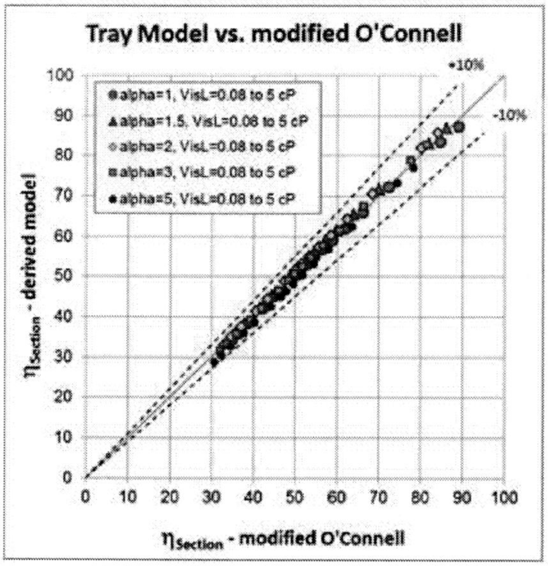

481687 Bubble Cap Tray Vapor Turndown

Abstract:

The UFM tray has been accepted globally as the next generation movable valve. It is only logical that this outstanding device be added to the Fixed Valve Arena for those applications that have a tendency to foul yet still need the high performance and high capacity that is provided by the UFM valve. This paper will show the capacity, performance and pressure drop of the new fixed UFM valve.

481688 Marinization of Mass Transfer Columns for FLNG Applications

Abstract:

Still by today, there is limited experience available about the impact of column motion on the fluid dynamic and mass transfer efficiency in industrial sized marinized columns. Raschig and UOP has started in 2010 a new approach to evaluate the performance of random and structured packings under column motion conditions. The experimental test results have been validated by CFD modeling. On the bases of CFD gas and liquid flow performance can be predicted at any elevated position mass transfer columns may have under combined heave, sway, surge, list, trim, roll, yaw and pitch conditions.

The paper offers new information about column operation for floating LNG applications.

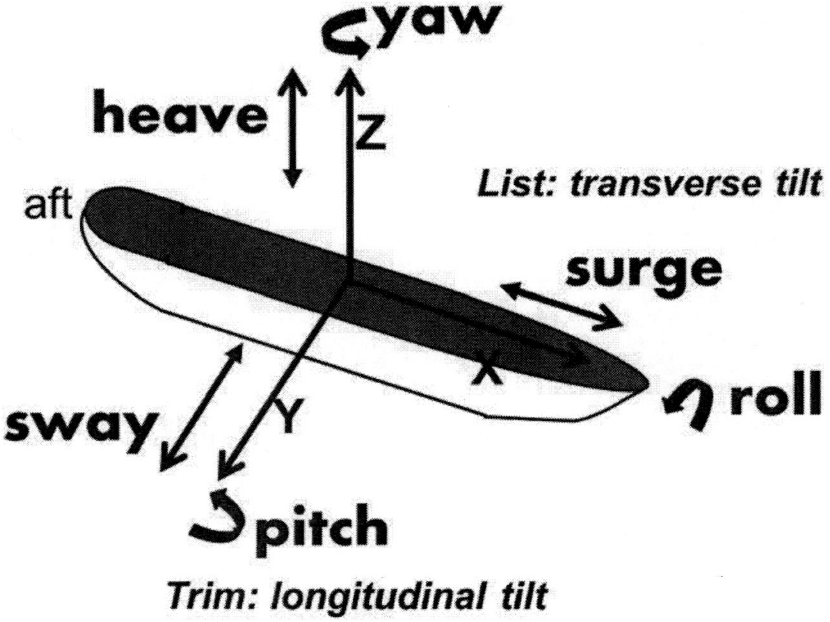

Fig. 1: Nomenclature for moving vessels

481689 Flow-Induced Tray Vibration and Its Prevention

Abstract:

In chemical processing industries (CPI), flow induced tray vibrations occurred with distillation and absorption columns using either sieve trays or valve trays. Numerous severe structural failures of industrial sieve or valve trays were attributed to flow induced vibrations. This paper discusses the basics of tray vibrations, and identifies tray parameters and tray hydraulics that affect the flow induced vibrations. Criteria and guidelines are proposed to prevent tray vibrations in new column designs or column revamps, and have been successfully applied to the industrial case of tray vibrations. The details of this industrial case and measures taken to prevent vibrations are presented and discussed.

479752 High Performance Trays for a Fractionator in an Aromatics Plant

Abstract:

Distillation is a very important separation technology in refineries and downstream chemical processing. Mass transfer internals in distillation columns can be equipped with trays, structured packing or random packing. Trays are generally the preferred option in columns with high liquid rates or in services with high fouling tendencies. After the acquisition of Nutter Engineering in 1998, Sulzer Chemtech expanded its tray product portfolio with inclusion of the acclaimed BDHTM rectangular float valves and the V-GridTM fixed valves. The technological alliance with Shell Global Solutions since 2000 has brought Shell high performance trays into Sulzer Chemtech's range of tray products. This case study highlights an integrated refinery complex in the East Asia, where various tray options were evaluated and Shell tray technology was eventually applied for a new xylene rerun column in an aromatic unit.

ADVANCE IN HIGH PERFORMANCE SPRAY TRAY DESIGN AND THEIR INDUSTRIAL APPLICATIONS

Zhuxiang Wang, Enxia Shang, Liming Zhang, Hong Ma, Tianjin ChuangJu Technology Co. Ltd, Tianjin, P. R. China

Yan Zhuang, Advanced Column Technology, Perth, Australia

Tianjin ChuangJu Technology Co., Ltd specializes in separation processes and equipment design, fabrication and installation, providing services to the chemical, oil and gas, pharmaceutical and environmental protection industries. ChuangJu developed its proprietary CJST series spray tray products and technologies with dozens of patents and know-hows, and successfully commercialised the spray tray technologies in nearly 1,800 industrial columns in 30 provinces of China and overseas, providing services to the chemical, petrochemical, oil and gas, coal chemical, methanol production, synthetic fibre production as well as pharmaceutical and other industries.

Key Words: spray tray, CJST, high performance, oil and gas, separation process

How does a Spray Tray Work

The spray tray utilizes a mechanical phase contactor, such as the vertical sieve cap as in New-VST spray tray, together with certain fluid dynamic conditions, to create a spray regime. In a vertical sieve cap contactor, the gas rises through the opening in the tray, while the liquid enters the cap through a gap between the cap and the tray floor. Inside the cap the liquid is drawn up, by the dynamic lifting of the rising vapor, as well as the liquid phase hydrostatic pressure on the tray. The liquid phase is broken into droplets, impinges to the top of the cap, which is the first stage of intensive phase contact; and then the vapor and liquid mixture is sprayed horizontally out through the vertical sieve holes, while the liquid droplets fall back to the tray and vapor rise up to the tray above, completing the second stage of phase contact. The liquid is the dispersed phase and the vapor is the continuous phase.

The spray type of contactor offers higher mass and energy transfer efficiency due to the following reasons:

- Increased total areas of liquid-gas interface by dispersing the liquid phase into a large amount of small droplets, which creates more interfacial contact areas;
- Longer phase contact time by two stages of phase contact ie inside the cap and outside of the cap;

- Faster interface renewal due to intensified phase contact, and
- Shorter diffusion path for liquid molecules within a small droplet compared with that in the bulk liquid.

Comparison of Conventional Tray vs Spray Tray

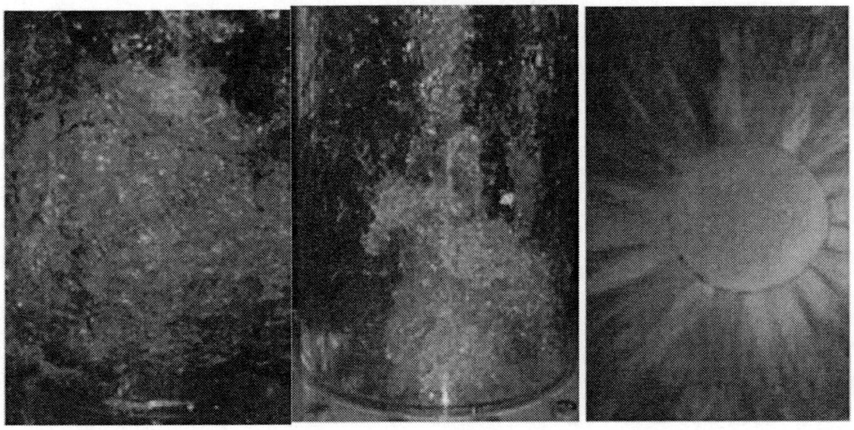

Froth Regime, Valve Tray **Spray Regime, Sieve Tray** **Spray Tray, New-VST**

Figure 1: Comparsion of Phase Contact of Different Tray Type and Operation Regimes

Column Tray Types

Conventional Trays	High Performance Spray Trays	
Bubble Cap	New-VST	
Sieve	CJST Series Spray Trays	CJST®
Valve		MPST Ⅰ ®
Dual Flow		MPST Ⅱ ®
		PFST®
		Low Back-mixing Spray Tray (LBST®)

New vertical sieve tray（New-VST）

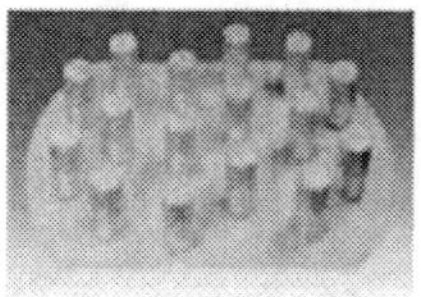

The New vertical sieve tray（New-VST） technology was developed by Japan's Mitsui Engineering & Shipbuilding (MES) in 1963-1968. In 1971, the New-VST was tested by the Fractionation Research Inc. and the results indicated higher mass transfer efficiency, higher capacity and lower pressure drop, compared with the conventional bubble cap, sieve or valve trays[1][2].

Since 1980s in China, the New–VST spray tray technology has been successfully applied to benzol stripping column, benzol refining column, ammonia distillation column, tar oil distillation column in coking industry, high boiling point distillation column and low boiling point distillation column of vinyl chloride rectification, saturated hot water column, ammonia distillation column, hybrid-carbonization column of the fertilizer industry, pre-rectifying column, pressure distillation, atmospheric distillation column, recovery column of methanol distillation, catalytic distillation etc[3].

Case Study for New-VST:

1、Coke Gas to Methanol

Jiantao Stage 2 Methanol Project Revamp using the New-VST Tray

The Jiantao Stage 2 Methanol Project was designed and supplied by another company using a float valve design. The atmospheric distillation column suffered several column internals break downs and the product quality failed to meet the design specifications. The facility could not reach design capacity. After investigation, ChuangJu concluded that the controlling step is the atmospheric distillation column. By replacing the float valve tray with the New-VST tray and retaining the original down comer design, ChuangJu solved the problem with rather low revamp cost. The revamped column passed the 24 hour test successfully.

Atmospheric Methanol Distillation Column	**Before Revamp**	**After Revamp**
Tray Type	Valve	New-VST
Methanol Production	16t/h	17.5t/h
Reflux Rate	17.9m^3/h	13m^3/h
Steam Consumption	1.25t/t product	1.0-1.1t/t product
Ethanol content in Methanol Product	<30ppm	<10ppm (meeting USA AA grade)
CH_3OH content in bottom discharge	0.4%	0.05%
Side withdraw of alcohol mixture	66.6t/h	0

2、 Methanol to Olifin Project

Inner Mongolia Fenghui Chemical Co., Ltd. Methanol to Olefins Project

The 100,000 tonnes / year Methanol to Gasoline Facility, built by the Inner Mongolia Fenghui Chemical Co., Ltd., started commissioning at 7pm on the 5th of November 2014, and produced on-spec gasoline (with RON up to 93) in the early morning on 7th of November. The facility production achieved 100% design capacity in less than 72 hours after starting up. The facility, consisted of fractional distillation unit include de-ethanizer, absorption tower, de-butanizer, and gasoline tower, was designed and manufactured by ChuangJu using the New-VST spray trays.

	Actual Operation
Tray Type	New-VST
Capacity	100,000 t/a
RON number	93
Bottom product of de-ethanizer C2-	≤2kg/h
Bottom product of De-butanizer C1-4 content	≤0.02%
Top product of De-butanizer C5+	≤2.7%
1234MBZ content light alkane stabilization column top product	≤0.5%
Single Tray Pressure Drop	400Pa

Radial side-guiding spray tray（CJST®）

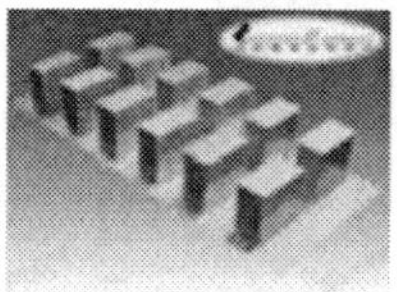

Radial side-guiding spray tray（CJST®） is an award-winning patented product and the patent number is ZL 2006 2 0025314.8. The product has since been successfully commercialized in coking, chlor-alkali, fertilizer, petrochemical and fine-chemical industries, etc. This tray design can handle high throughput. The column pressure is lower compared with a valve tray design, is fouling resistant, and especially suitable for large diameter columns. The liquid is spayed out toward the downcomer direction which reduces the liquid back-mixing tendency.

Case Study for CJST®: Atmospheric and Vacuum Oil Distillation Columns Xi'an Petrochemical Ltd. Sinopec Corp

The capacity of Xi'an Petrochemical's atmospheric distillation column was 700,000t/a. Due to the market requirement the capacity of the unit needed to increase to 1,100,000t/a.

The original column used a valve tray. When the throughput increased to 800,000t/a, the column separation efficiency sharply dropped. In 2011, the column was revamped using ChuangJu's CJST® tray. After the revamp, the throughput increased to 1,200,000 t/a, a 71% increase compared with that before the revamp. Moreover the column pressure drop was significantly reduced from a maximum of 55kPa to 25kPa.

2、 Natural Gas Industry

No.1, 2 and 3 Natural Gas Treatment Plants of Changqing Oil Field of PetroChina

This project was one part of China's West to East Natural Gas Transmission Project. The natural gas sweeting plants all were imported from overseas. The columns used floating valve trays by Kohn-Glitch. During start-up, the following problems occurred:

- Severe foaming caused flooding;
- The amine loss was high;
- The throughput is lower than design capacity;
- The gas output quality fluctuates;
- Severe corrosion problems in pipework, and
- Poor product quality in the sulfur recovery unit.

- From 2007, ChuangJu revamped all the columns in the three gas sweeting plants with its CJST® trays with successful results.

Gas Sweeting Column	Before Revamp	After Revamp
Column tray design	Kohn-Glitch floating valve	CJST® spray tray
Processing throughput	1.5 million-2.3million m^3/d	3million m^3/d
Single Tray Pressure Drop	1700Pa	750-900Pa
Steam Consumption	1.08t steam/10,000 m^3	0.86t steam/10,000m^3
Flooding occurrence/month	3 times	0
Anti-foaming Agent Dosage	1000-3000mL/d	0

Low resistance membrane spray tray（MPST I ®)

The low resistance membrane spray tray (MPST I) patent number is ZL200620152162.8. This tray was developed for using in the ammonia recovery column from the copper wash waste water in the ammonia fertilizer industry.

<6>Membrane spray without back-mixing tray（MPST II ®）

This technology is a new product developed by modifying the above mentioned (MP I ®) to solve the problem of liquid back-mixing on the tray. It is suitable for process conditions with high liquid-vapor ratio or fouling susceptible systems. In this design the liquid that has been contacted in the membrane spray contactors is separated from the fresh liquid from the tray above, thus eliminated the problem of liquid back mixing on top of the tray.

Main characteristics:

（1） No liquid back-mixing and concentration gradient on the tray,

（2） High mass transfer efficiency,

（3） Low pressure drop. The tray resistance can be controlled to below 20mmH_2O.

（4） Strong anti-fouling capability

（5） Suitable for vacuum distillation or vacuum absorption, and absorption with heat exchange requirements.

Case Study: Qilu Petrochemical Technology Ltd 's Wastewater Stripping Column

The original waste water strip column was a DN2000, with 44 valve trays, designed to treat 80-120m3 waste water containing S components at 17,000 mg/L and NH_3-N at 12,000mg/L.

Before the revamping, the valves were blocked within one or two months of operation, and the treated waste water did not meet the design specification. ChuangJu revamped the column using its CJMP II and solved the problem for the Client. The results before and after revamp are summarized in the table below. .

	Before Revamp	**After Revamp**
Tray Type	Floating Valve	MP II
Processing Throughput	110m^3/h	140m^3/h
Ammonia Content	300 -400mg/L	80mg/L
Steam Consumption saving		10-15%
Maintenance shutdown	Once per month	Once per 12 month

2、Amine Regeneration Columns

Qilu Petrochemical Refinery Ltd.

Qilu Petrochemical Refinery's FCC Unit Amine Regeneration Column (DN2800) had been suffering flooding and liquid entrainment since October 2012, which limited the column's throughput to only 210m^3/h. After replacing the float valve trays with ChuangJu's MP trays, the throughput has been increased to 240m^3/h with steady operation since May 2013.

	Before Revamp	After Revamp
Column Internals	Float Valve	MP II
Column Throughput	210m^3/h	240m^3/h
S content in Regenerated Solvent	N/A	≤1.6g/L
Steam Consumption	N/A	<16.5t/h
Single Tray Pressure Drop	2480Pa	＜440Pa

3、Water Treatment

Shaanxi Coal Chemical Group Shenmu Tianyuan Chemical Co., Ltd. 100t / h waste water treatment project

High phenol wastewater, called semi-coke wastewater, comes from coal distillation units and crude aromatic hydrogen production unit, which contains high concentrations of oil, organic components with phenol, ammonia nitrogen et.al. The semi-coke waste water contains much higher COD content compare with wastewater generated from other coking plant, especially the oil content can be up to 10,000ppm with the majority as emulsified oil. The traditional process is no longer applicable to the treatment of semi-coke wastewater.

ChuangJu developed a process specifically for the treatment of the semi-coke wastewater using its proprietary CJST® anti-fouling spray tray technologies. The process can effectively improve the efficiency of equipment, reduce the amount of emulsified oil, and recover ammonia in the wastewater to produce anhydrous ammonia with 99.9% purity, as well as oil, phenol and sulfur in the waste water.

	Design Value	Achieved Value
Column Internals	MP II	MP II
COD concentration	≤3500mg/L	2300mg/L
NH3-N	≤200mg/L	100mg/L
pH	5-7	7
Volatile phenol	≤400mg/L	220 mg/L
Oil content	≤180mg/L	150 mg/L
Sulfide content	≤80mg/L	50 mg/L

ChuangJu has successfully design or revamped over 70 columns using its own proprietary MP series spray tray technologies in the Coal Chemical Industry since 2009.

Partitioned Flow Spray Tray（PFST®）

The PFST® tray is designed with a sandwich plate to catch the liquid that has been sprayed out of the vertical sieve caps after mixing with the vapor. The liquid is then channeled down to the next tray by a multiple down comer design to handle high liquid load, thus eliminates the problem of liquid back mixing on the tray. With this design, the vapor and liquid phases come to contact counter-currently with the maximum mass transfer driving force, everywhere on the tray, without any liquid flow and concentration gradients. The spray action of the vertical sieve cap reduces the foaming tendency, thus reducing the tendency of vapor phase back mixing.

Case Study for PFST®：

Raw C4 Downstream Processing

1、Luoyang Petrochemical Hongda Industry's Raw C4 Downstream Processing (100×10^3 t/a)

In September 2013, the Louyang Petrochemical Hongda Industry Ltd started its 100,000t/a Raw

C4 Downstream Processing Facility. The facility includes an extractive distillation, a de-gassing and a solvent regeneration column. ChuangJu designed and manufactured all the three columns. The extractive distillation column and the de-gassing column used ChuangJu's PFST® tray. The facility reached the design capacity and met or exceeded all products specification within 65 hours of start-up with steady operation.

Parameters		Design Value	Actual Operational
Temperature℃	Extractive Distillation	132	118.8
	De-gassing	172.9	165
Steam Consumption kg/h	Extractive Distillation	3,117	2,700
	De-gassing	5840	5000
Isobutane Product Concentration %		≥96%	98%
Olefins Concentration %		≥60%	66%

2、**Ningbo Haode Chemical Industry Ltd.**

Ningbo Haode Chemical Industry Ltd. owns and operates a facility that produce 47,000t/a of isobutene. The facility consists of a C4 separation, n-butane isomerization unit, a MTBE synthesizing unit and a MTBE cracking to isobutene unit. The capacity of the C4 separation unit is 120,000t/a, using a solvent system of MEK and NFM. This system uses a high liquid to vapor load and is susceptible to foaming.

The original design of the columns used a fixed valve tray. After the start up in March 2016, the unit could only run at 70% of the design capacity. ChuangJu was invited to investigate and solve the problem. After site visit with ChuangJu's experienced technical personnel, they found the issues included: the liquid flow cross section is too small; low the tray separation efficiency is low, and there is severe foaming which makes the column easily flooded.

ChuangJu chose their PFST® tray to revamp the two extractive distillation columns and the solvent degassing column. The revamped columns were started up successfully in June 2016.

The achieved throughput is increased by 70% compared with that before the revamp, and has been running at 30% above its design capacity since then.

Parameters		Before Revamp	After Revamp
Type of tray		Fixed Valves	PFST®
Feed rate, kg/h	**Distillation Columns A/B**	12,560	21,600
	De-gassing Column	124,660	226,930
Temperature℃	**Distillation Columns A/B**	142.8	140.3
	De-gassing Column	156.8	154.4
Steam Consumption kg/t feed	**Distillation Columns A/B**	520	295
	De-gassing Column	53.8	37.5
Solvent entrainment, ppm	**Distillation Columns A/B**	1,000	0
	De-gassing Column	1,000	0
Product Butane Content %		96%	96%
Total olefin content%		90%	90%
Scheduled maintenance times per year		4	1

Low Back-mixing Spray Tray (LBST®)

In 2015, in order to increase mass transfer efficiency and reduce liquid phase back mixing, ChuangJu developed the Low Back-mixing Spray Tray (LBST®) as an improvement to the New-VST technology. The LBST® tray modified the down comer design and added a liquid distribution trough to feed fresh liquid to each individual vertical sieve caps on the tray, so each cap will have the same mass transfer driving force to achieve a uniform efficiency. The patent No is 201521131476.5.

The main advantages of the LBST® tray:

- **Uniform and high mass transfer efficiency**: With its unique down comer and liquid distribution through design each of the vertical sieve cap is surrounded by an equal level layer of fresh liquid from the tray above, so each of the caps will have the same high mass transfer efficiency, thus a high tray efficiency.
- **Low liquid phase back-mixing**: Due to the unique down comer design, most of the liquid that sprayed out of the caps is separated from the fresh liquid from tray above, so the liquid back-mixing is very low.
- **Wider Operation window**: The liquid and vapor contact remains the same when the liquid flow rate varies due to the unique down comer design.

Case Study for LBST® Tray :

Coal to Methanol Technology

Revamping of the Pressure Distillation Column of ENN's 600,000 t/a Coal to Methanol Plant

At the request of ENN Xinneng Energy, ChuangJu was asked to modify the pressure distillation column which could not reach the design throughput. The pressure distillation column used valve trays, which could not handle the required liquid load and run into flooding and poor separation efficiency, especially at the rectification section. After investigation and modolling, ChuangJu decided only to replace the 10 trays in the rectification section with their LBST® spray trays, thus improve the whole column (has 69 trays in total) operation and solved the problem.

	Before Revamp	After Revamp
Tray Type	Valve	LBST®
Pressure distillation column feed rate	$90 m^3/h$	$130\ m^3/h$
Ethanol content in Methanol Product	600ppm	<100ppm

References:

[1] Li, Zhiqiang, Xu, Qingmao, etc., New-VST Technology, "The Journal of Chemical Industry Development", 2001, 20 (10): 58-60

[2] Du, Peiheng, Liu, Jincheng, The development and review of New-VST, "The Journal of Petrochemical Equipment", 1988, 17 (1): 27-31

[3] Du Jianting, Du Peiheng, The Fluid Dynamics and Mass Transfer Properties' Study on New-VST in China, "The Journal of Petrochemical Equipment", 2002, 31 (5): 35-40

480575 Packed Bed Performance Analytics Based on Gamma Scans

Abstract:

Refinery and chemical plant operations depend heavily on distillation and separation towers. Tower gamma scanning is well established in the process industries as a qualitative tool to help troubleshoot towers. Advancements in data analysis have led to a quantitative approach in expressing gamma scan data in numerical terms easily understood by process and operations engineers.

For packed towers, a grid-scan of 3 or 4 equal-distant scans crossing through the beds of packing would typically be used to investigate the quality of liquid distribution. The conventional approach to "analyzing" a gamma scan has been to visualize how well the scan data from the individual scans matched each other or how well they "overlaid" with each other. This is a totally subjective analysis lacking consistency, open to varying interpretation and does not translate well from tower-to-tower. Therefore the resulting conclusions from this approach can be very ambiguous regarding magnitude of any detected liquid mal-distribution.

An alternative analytical approach, termed PackView™, has been developed whereby data from a grid-scan provides a relative density scale. The density scale begins at the density of the dry or non-operating packing. To derive this value it is necessary to know the packing type to reference its dry bulk density. The density scale displays the calculated density of liquid retained in the bed of packing based on the scan results. As with typical gamma scan analysis, if the four scanlines have matching liquid retention densities then the implication is that the liquid distribution is good. However, if there is a difference between the scanlines, the retention density gives a numerical comparison from which to gauge the extent or severity of any liquid mal-distribution.

Another calculation by which to put the liquid distribution into perspective and to get a measure on the useful capacity of the packing is to calculate the liquid holdup fraction or liquid volume fraction. If the measured liquid retention density is divided by the process liquid density at bed conditions (the liquid density at the actual operating temperature and pressure), liquid holdup or liquid volume fraction can be established. A comparison to packing operating capacity curves allows this fraction to provide an objective appraisal of current operating capacity.

It is always easier to understand and discuss technical issues when quantitative information can be used to compare operational parameters with engineering design. Over the past 40 years gamma scanning has become more and more popular as a useful diagnostic tool to understand the hydraulic operation of fractionation equipment. This advanced analysis provides a new method of extracting quantitative information from gamma scan data to diagnose and characterize operation of distillation and separation towers. It is our goal that using the advanced analysis presented will improve the value of gamma scan data and facilitate improvements in the operation of mass transfer equipment.

Rectifying a Reactive Distillation Column

By Greg Theunick, The Dow Chemical Company, Midland, MI

ABSTRACT

Reactive distillation technology is used to produce Material A silane. Since the inception of making this material at a commercial scale, the process has had issues challenging stable operation. A troubleshooting investigation solved long term issues of pressure instability, steam surges and achieving product specifications within a limited startup timeframe. This paper captures the diagnosis and resolution for each issue.

BACKGROUND

"Material A" silane is made using reactive distillation technology. The reactive column is a shared asset which gets utilized for several reactive distillation products. Since Material A has been made at a commercial scale it has had issues that have made production a challenge. Other products made on this asset have experienced similar issues but Material A has been the most impacted.

Material A is produced by reacting Material B and C (see equation 1 below). A noncondensible coproduct is also formed. This is an equilibrium limited reaction so the products are quickly separated to keep the reaction favoring product formation. Some of Material B is in the liquid phase and co-fed to the column with Material C. This ensures good mixing of reactant species before the material enters the column. A fraction of Material B is vaporized and introduced lower in the column to reduce the amount of unreacted material leaving in the bottoms product. Material B is a low boiler relative to the bottoms product so it quickly migrates up the column and reacts with any unreacted material it encounters along the way. Only noncondensibles vent out, all other materials come out in the bottoms product. This reaction has the potential to form undesired side products. These are impurities that require additional processing downstream before the final product can be sold.

Eq. 1: *Material B + Material C* $\longleftrightarrow$ *Material A + Noncondensible*

Figure 1 is a high level schematic of the distillation asset. The process equipment includes:

- Column 1 – This is the reaction tower and contains structured packing internals. Liquid feeds are introduced at the top of the column and the vapor feed is introduced above the bottom packed bed. The reboiler operates as a thermosiphon with steam as the heat source for the exchanger. On the overhead there are two condensers. Reflux gravity drains from the second condenser (column 2) to column 1.

- Condenser and Vapor/Liquid Separator - The first condenser is a vertical shell and tube unit connected to a vertical vapor/liquid separator. Process material flows through the shell side and a refrigerated liquid is on the tube side. In the condenser the refrigerated material does not contact the process material. As the refrigerated material vaporizes, it provides cooling to the process. The tube side is directed to the vapor/liquid separator which ensures only vapor is returned to the refrigerated material supply process. The shell side outlet is a two phase stream that continues into the next condenser (column 2).

- Column 2 - This packed column functions as the second condenser. In this column the same refrigerated liquid is introduced as was used on the condenser but in this vessel the liquid contacts the vapor rising through the column. This creates the functionality of a contact condenser.

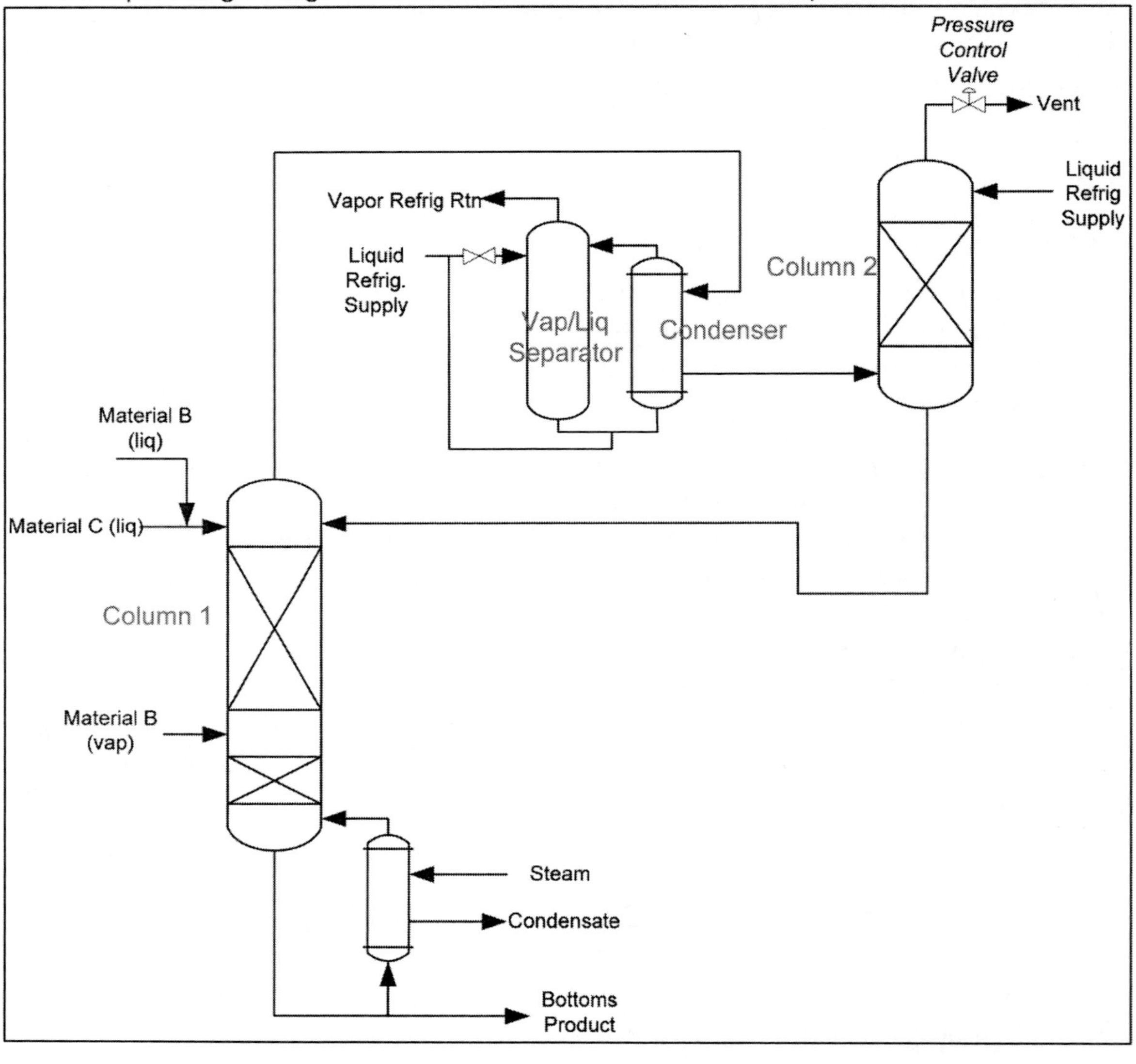

Figure 1: Process Diagram for reactive distillation asset and supporting equipment.

There are three main challenges impacting Material A production; unstable column pressure, steam flow surges and an inability to achieve product specifications in a short startup window. The following sections provide details on the troubleshooting and resolution for each of these issues.

PRESSURE INSTABILITY

Stable column pressure is an essential part of stable distillation operation. Pressure impacts column temperatures which in this application are critical as a way to monitor column performance and reaction conditions. Pressure also impacts hydraulic loading, vaporization, and condensation.

Often while producing Material A (and some other products), the column pressure would cycle and not be controllable with the pressure control scheme. Figure 2 shows plant operating data during the pressure cycling event. The cycling does not occur every campaign but when it does the cycle has a consistent repeating pattern with peaks approximately every two minutes. With this cycling the column temperatures would also cycle and become unreliable to use as process performance indicators. At times the cycling was so extreme that the pressure would drop low enough to allow the sump to flash, rapidly shifting column inventory and loading up the packing. These were significant upsets and it would take time to get the column lined back out, sometimes even resulting in a shutdown.

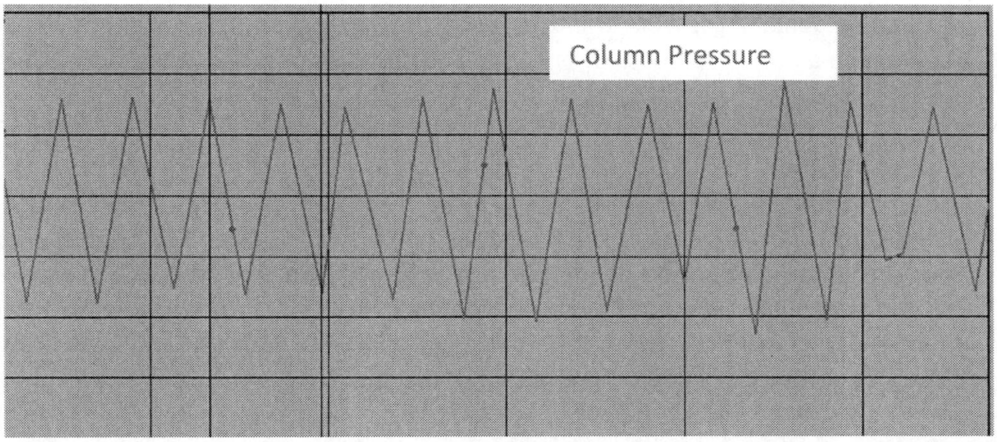

Figure 2: Column overhead pressure readings during cycling period. Peaks are approximately every 2 minutes, consistently, swings 20% of operating pressure

As this problem has persisted for several years many theories have been developed as to why this pressure cycling occurs. The most credible theories were investigated further:

- Tuning Problems – theory was that poor tuning was causing pressure instability. Improper tuning of control loops has been known to result in similar problems controlling pressure. The team conducted trials to develop a new set of tuning parameters for the column pressure control. However the pressure oscillations persisted after the tuning updates were made so tuning was not believed to be cause of the problem. The pressure control valve was even put in manual for a period of time which had no impact on the pressure cycling, which was also a clue something else was dominating the pressure control.

- Mid-boiler Accumulation – Other reactive distillation columns on site experience mid-boiler accumulation with the extreme temperature difference between the condenser and the column bottoms. This tends to trap in mid-boiling species that accumulate over time and have been known to cause column issues. However the problems with this column appear from the start, prior to allowing a mid-boiler to accumulate to a sufficient quantity to cause issues. This theory was dismissed.

- Column 2 Inlet Flooding – The reflux return to Column 1 is approximately 2ft elevation lower than the inlet to the second condenser (Column 2). The theory was the pressure drop in the overheads was causing a higher liquid head in Column 2 in order to gravity drain reflux back to Column 1. If the inlet to Column 2 was flooded, vapor would bubble through the liquid causing erratic differential pressure in Column 2 as the vapor carries liquid up into the packing. This would impact the overall system pressure. Column 2 and the gravity drain line were gamma scanned during stable operation and during a pressure upset. In both cases Column 2 inlet was not flooded. Therefore this theory was dismissed as the possible cause.

- Two phase flow in lines (surging) – The outlet of the condenser has both liquid and vapor flowing in a horizontal section of pipe. The theory was the line was experiencing slugging flow which was causing periodic surging and pressure instability. After investigating further it was found that the line was adequately sized for this two phase flow. Plus the pressure oscillations showed an even amount of time building pressure and then losing pressure. In a surging event, it would have been expected to see a more rapid loss of pressure and a slower increase in pressure. This theory was dismissed as well.

Fixing the Pressure Problem

Working with the operators and reviewing the historical plant operating data, it was discovered that when the temperature setpoint between the condenser and column 2 was lowered, the pressure oscillation stopped. Figure 3 shows plant data and how the pressure steadies as the temperature is lowered. The methodology works every time, but the setpoint needed varied each time. This worked for the other products that were also experiencing the same pressure instability.

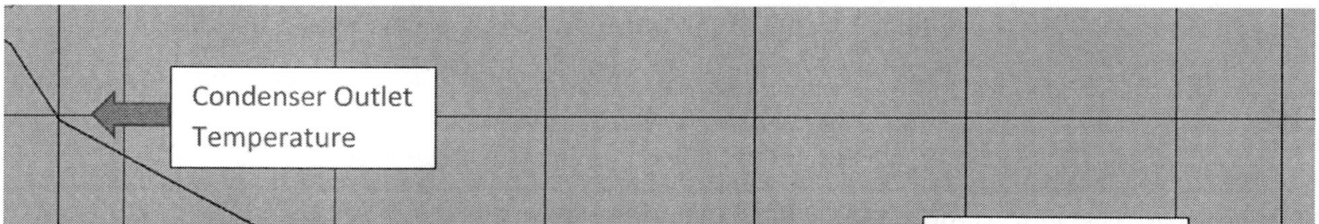

Figure 3: Plant data showing as the condenser outlet is cooled (blue) the column pressure becomes stable (red).

What Was Fixed?

A methodology was developed that solved the problem but the root cause was still unknown. The process seemed sensitive to the condenser outlet temperature but what was changing? Was this even the right variable? To validate that this was the correct variable a trial was conducted going the opposite direction. If cooling down the temperature stabilized the pressure, then warming it up could destabilize things. The condenser outlet temperature was warmed and as predicted the column pressure cycling returned as shown in Figure 4.

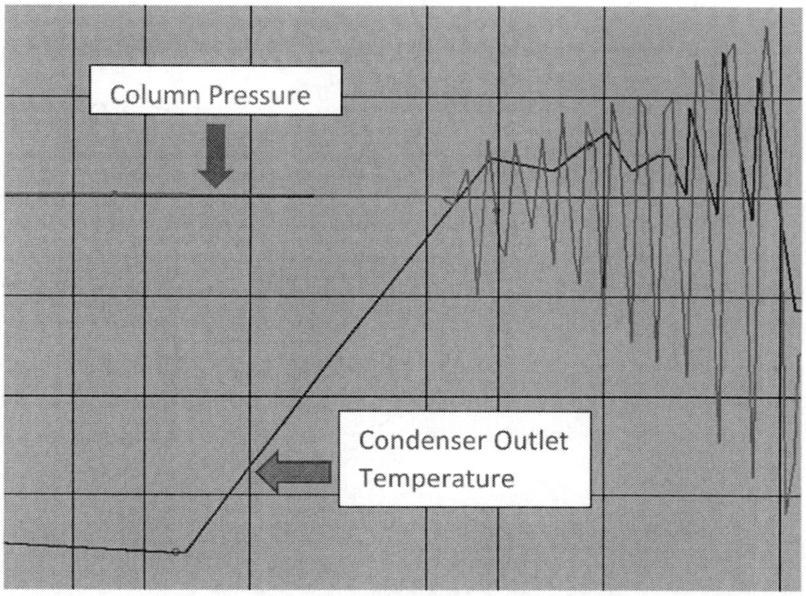

Figure 4: Plant data showing as the condenser outlet temperature increases (blue) the column pressure becomes unstable (red).

Further investigation revealed that the pressure oscillations occur only when condenser was used. Some products bypass the condenser and only use Column 2 to condense the overheads. This caused a more focused look at the condenser and its companion vessel. It was concluded that condenser was meant to operate as a thermosiphon evaporator. A thermosiphon is commonly used as a reboiler with process material on the tube side and steam or some other heat source on the shell side. For this reactive tower, the process is on the shell side and refrigerated liquid is on the tube side. The vapor/liquid separator functions similar to the sump of a column, providing vapor/liquid separation, residence time and ability to provide static head to drive the thermosiphon.

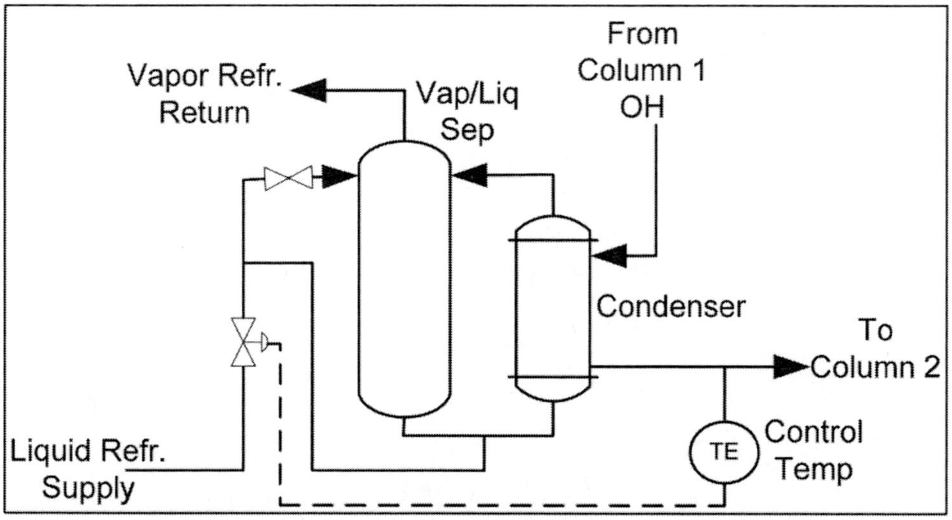

Figure 5: Condenser and Vapor/Separator functioning as a thermosiphon.

With the condenser functionality discovery a connection was still needed explaining how a thermosiphon could be related to the pressure cycling. Leveraging a troubleshooting effort from a separate process, a thermosiphon was unstable and as a result the shell side pressure was also very erratic (see figure 6). Once the thermosiphon vaporization was stabilized the shell side pressure also becomes stable. This thermosiphon was fixed by increasing the sump level so there was sufficient head to drive the thermosiphon circulation.

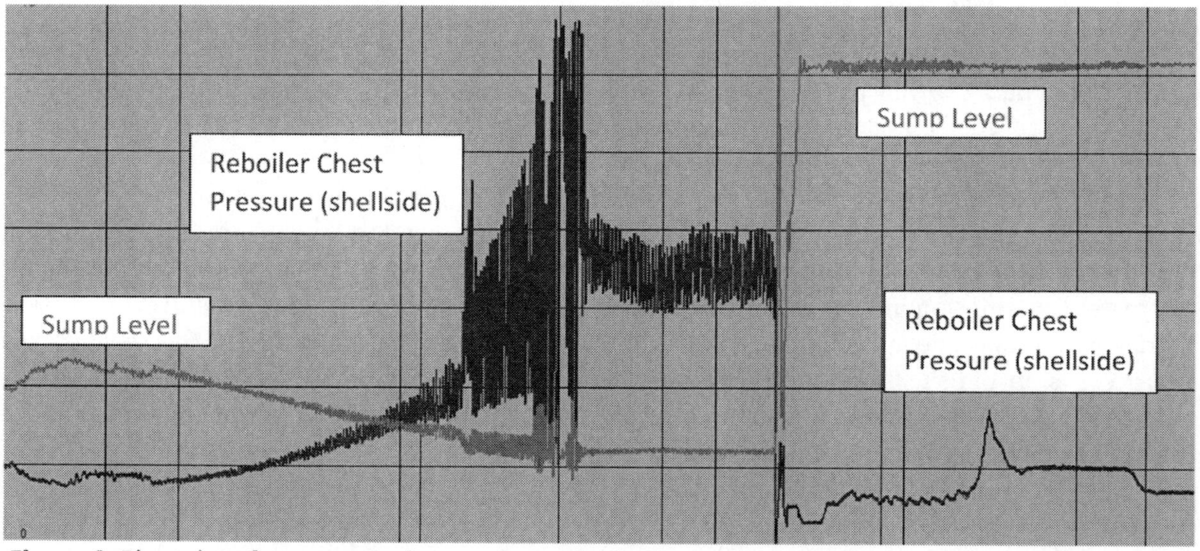

Figure 6: Plant data from a typical steam heated reboiler on a column. Chest pressure (shell side) is shown in blue, sump level is in red. Pressure oscillations stop once level is increased.

The previous example helps put the pieces together to explain why the methodology used to fix the pressure actually works. For Material A production, when the operators implement the solution of decreasing the temperature setpoint on the condenser outlet, the control system responds by increasing refrigerant flow. With more flow the level will increase in the vapor/liquid separator (which acts like a column sump in this installation). This increased level provides head pressure allowing for a more stable thermosiphon and the result is stable column pressure which is on the shell side of the unit. Just like the separate process example shared above, more level in the "sump" fixed the problem and steadied the shell side pressure, in this case the column pressure. Gamma scans on the vapor liquid separator show the level increase in the unit as the temperature setpoint is lowered, which helps validate the theory.

With this new understanding of what causes the pressure instability and the already established methodology to stabilize the column pressure; the process has been able to successfully run with a controllable pressure.

STEAM SURGES

Steam is supplied to the reboiler on the column as the heating medium. At times, the steam flow to the reboiler experienced sudden surges in flow up to 150% of the control setpoint. This would be followed by a loss in steam flow to the reboiler. In the process history steam surges have caused a number of upsets and have affected Material A production.

The issue wasn't unique to Material A, other products made on the column also had issues related to steam. While making another product, the process could only run a few days before the steam valve would open to 100% and flow would drop off. Precedence from previous troubleshooting experience led to the theory that this was an indication of the reboiler flooding with condensate.

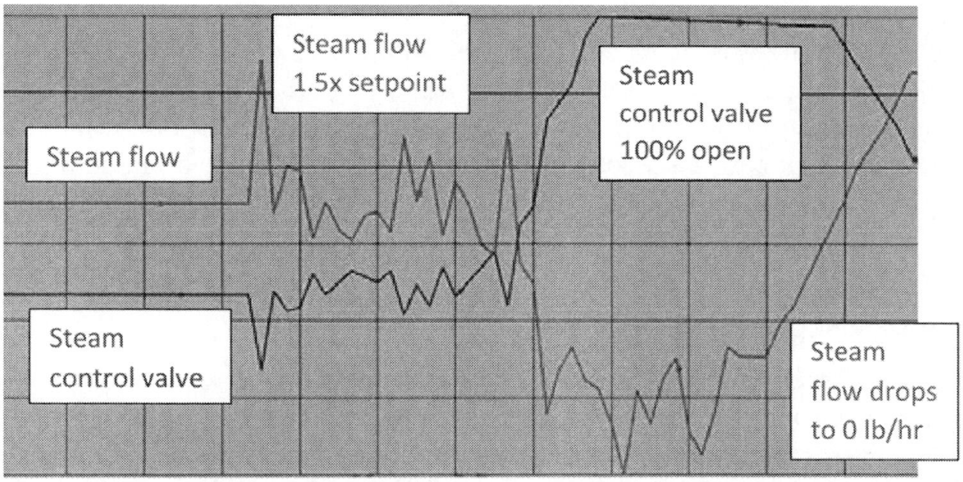

Figure 7: Plant Data showing steam flow (red) and steam control valve position (blue). Steam flow surged above setpoint. Then flow dropped off while steam flow valve went 100% open.

The steam line from the highline to the reboiler was walked down. It was found that there were areas of the existing installation that were not best practice. The main takeoff from the highline came off the bottom of the header, which made the branch vulnerable to condensate from the highline. Then as the piping worked its way through the control valve station and out to the reboiler, there were two large sections of pipe with low points and no way to drain any condensate. These pockets could collect condensate, which could eventually slug to the reboiler flooding in the reboiler shell. Even around the control valve there was a pocket, which could collect condensate and create a false flow measurement. Figure 8 shows a representation of the condensate collection points in the piping. This theory matched up well with all the evidence and historical plant data.

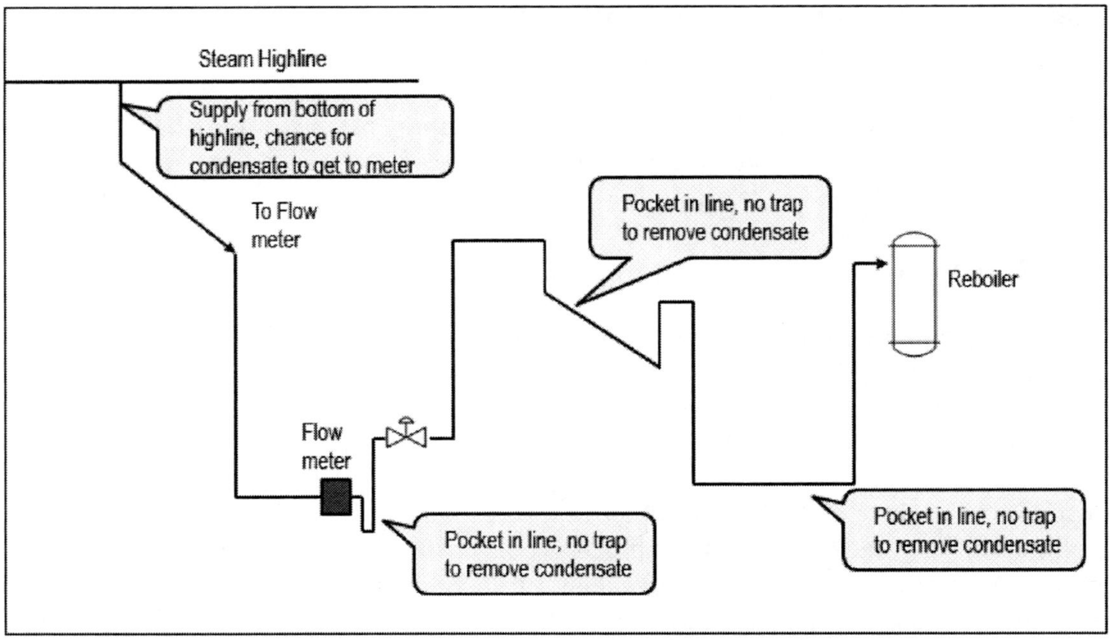

Figure 8: Orthographic representation of steam piping from highline to the reboiler.

Once trapped condensate was identified as the likely root cause, a temporary solution was put in place by adding drip traps at the low points in the steam piping. This gave any condensate a way out without impacting the reboiler. Later a capital project repiped the steam line to the reboiler to eliminate the pockets. Since the solutions have been put in place, there have been no reoccurrences of the steam surges.

OUT OF SPEC STARTUP MATERIAL

At startup all out of spec material leaving the column bottoms is segregated to a collection tank, which has limited capacity. Once the tank is full if the column bottoms is still out of spec the process has to shutdown. Shutdown is followed by a series of cleanout steps to essentially "reset" the column for another try at making Material A. This generates a lot of waste and consumes valuable raw materials. The last remaining challenge impacting successful operation relates to getting product that meets specification before exceeding the startup tank volume.

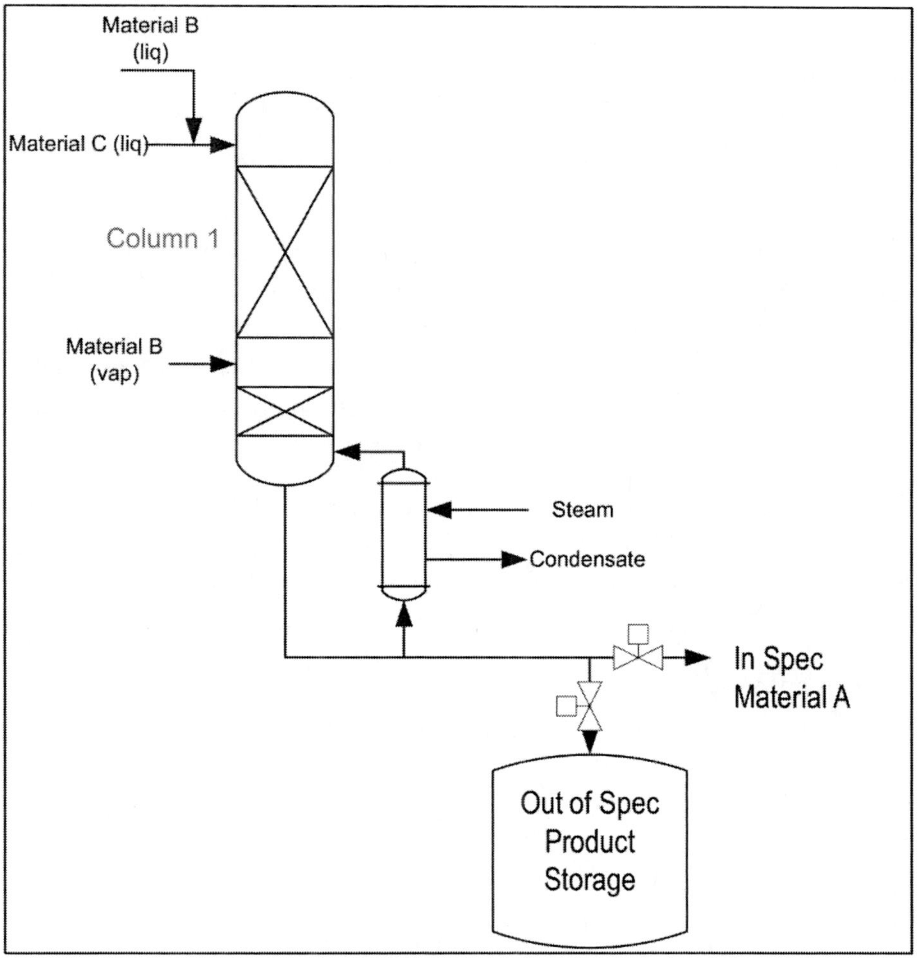

Figure 9: Bottoms Product diverted to holding tank until it meets specification

The startup material is out of spec due to the way the process is intentionally started. To avoid creation of unwanted byproducts, the process is started with less than the stoichiometric amount of Material B needed to complete the reaction. As a result, Material C is only partially reacted which causes the bottoms to be out of spec high on this material. While this is effective at managing side product formation it creates an inventory of out of spec material that must be corrected.

Historically to recover from this condition, more Material B is added in small slow increments above what is needed for the stoichiometric ratio. This recovery period takes most of the startup duration. Sometimes the product cannot be brought within spec by the time the out of spec tank fills.

Since the current solution is to slowly add more material B, a theory was developed to add Material B at a faster rate to see if the process would recover more quickly. Plant trials were completed and it was found that Material A came into spec much faster with this approach. The addition rate was optimized to the point where a 50% reduction in the startup duration was achieved. This addressed the problem and resulted in financial benefits for the process.

EPILOGUE

With the solutions implemented the plant has successfully run several dozen stable campaigns over the last year without any reoccurrence of the previous issues. Better still is the added process control and stability has allowed for extended campaigns achieving higher production volumes.

This experience sheds light on the unique application of a thermosiphon exchanger functioning as an overhead condenser. The paper has shown the challenges that can occur if this type of exchanger is not functioning properly and the impact that can have on the process. For years the production facility operated without understanding how the condenser was designed to operate and as a result went through several unsuccessful attempts to resolve the pressure instability with no effect. However once the true intent of the exchanger was understood and a methodology for control was established, the process was able to develop a robust pressure control strategy.

This study also emphasized the value in walking down a process to look for installations that are out of the ordinary or not best practice that could be causes for process problems. A simple walk down revealed the cause of the steam issue. This is a valuable lesson that valuable troubleshooting knowledge can be gained by a trip out to the process.

Finally as encouragement to future troubleshooters, don't be afraid to step up to a challenge. It doesn't matter how many have come before you and have been unsuccessful, all it takes is the right set of experiences to know how to solve a problem.

REFERENCES

1. Kister, H.Z., "Distillation Operation", McGraw-Hill, NY, 1990

2. Kister, H.Z., "Distillation Design", McGraw-Hill, NY, 1992

Distillation Control - Still an Issue

Sigifredo Nino, P.Eng.

Summa Control Solutions Inc.

Montreal, QC, CANADA

SNino@SummaControlSolutions.com

Presented at the 2017 Distillation Symposium, AIChE Spring Meeting, San Antonio, Texas, March 26 - 30, 2017

DISTILLATION CONTROL – STILL AN ISSUE

Sigifredo Nino, Summa Control Solutions Inc., Quebec, Canada
Email: SNino@SummaControlSolutions.com

Distillation control is and will always be a challenging application for the process control engineer. While not a subject of active research, its relevance remains. Moreover, being a multivariable control application does not make it a perfect fit for Model Predictive Control either. The distillation columns are typically delivered to the end user with the conventional LV controls where reflux (L) and boilup (V) control top and bottom composition respectively. The bigger the reflux ratio, the harder it becomes to control the overhead drum level; it will take an enthalpy change for this dual-composition controls array to interact. That is to say, a change in the reflux flow and the subsequent change in reboiler heat input will result in an oscillation that will only be stopped by operating a single composition in closed loop and leaving the other manipulated variable for the operator, to keep the product composition specifications in check and providing the slack to deal with unmeasured disturbances. The end result is the over purification of the column main product by using more energy than required, and the missed opportunity to optimize product recovery.

When it comes to distillation control strategies, there is no "one size fits all" approach. A clear understanding of the column's behaviour and purpose are critical to the design of a control strategy that can respond to specific requirements. For this reason, distillation control remains, and will likely always be, an issue. Due to the disappearing understanding of Advanced Regulatory Control (ARC), the use of Model Predictive Control (MPC) is currently being favoured, according to McMillan et al. [4]. However, as stated in this article, in which the author collaborated, how can the deficient understanding of a process be superseded by the use of an MPC tool that is not properly comprehended?

None of this should come as news. As the pendulum swings back to remind the industry that it is time to optimize production units, the purpose of this paper is to demonstrate through case studies how material balance controls have been implemented to improve/stabilize several distillation columns. These improvements have invariably resulted in quantifiable economic benefits, with an atypically short timeline for Return On Investment (ROI) measured in days, not months.

Acknowledging MPC's place in process control and understanding that Process Control technologies are meant to complement each other and not to be in competition, one of the presented cases studies is specific on the selection of ARC over MPC.

For this article, a couple of notes are in order: First, since temperature sensors located within the column can be used as composition surrogates, the text may refer to temperatures when mentioning a product composition. Secondly, the distillation column models described in this paper are only suitable for process analysis and control strategy design.

CASE 1: Xylene Tower: Single composition control means one degree of freedom available for optimization

This column is the last in a BTX aromatics train at an oil refinery. It was designed to separate the petrochemical feedstock exiting as distillate product from the ortho-xylene to be removed as vapour sidedraw below the feed tray and the bottom stream directed to the gasoline pool. The bottom sidestream was no longer in use when the ARC optimization was implemented. In consequence, the ortho-xylene distributed along the tower exits with both distillate and bottoms product.

The petrochemical feedstock coming out as distillate has more value than the bottom product. Since there was no specification for the bottom product, it was an opportunity to optimize the distillation operation profitability by bringing the top product non-aromatics impurities to the allowed maximum.

In the original control strategy, top composition was not controlled and bottom composition control was attempted by using the fired heater reboiler Coil Outlet Temperature (COT) with little success, as this temperature is not a reflection of the components in the tower bottoms. The reflux drum level was controlled simultaneously by the reflux and the distillate flow by using a reflux ratio (RR) set up by the operator.

Shinskey's four-component model operating lines and relative gains methodology [8] was used to analyze the possible selections of a single-end composition control pairing, under the scenarios of the predominant operating conditions, and to compare them to the optimization mode that would maximize bottom product recovery into the distillate, as it is the most valuable stream. For completeness sake, that methodology is briefly outlined herein; interested readers will find the details in [7].

The starting point of the methodology is the use of relative gain (RG) as the quantification of the interaction amongst control loops in a multivariable system. Relative Gain metric was proposed by Bristol [1], and its application to a Two Inputs Two Outputs (TITO) multivariable system is a perfect fit for its use in the selection of distillation control configuration.

In a two-loops interacting process control system, there are 2 controlled variables (CV1 and CV2) and 2 manipulated variables (MV1 and MV2). If we designate the difference of a variable between final and initial steady state values by Δ, and the relative gain by λ, the latter is calculated by the ratio of the change in CV1 caused by a change in MV1, when both loops are in manual, over the change in CV1 over the MV1 change when the second control loop is in automatic, as shown in Eq. 1

$$\lambda = \frac{(\Delta CV_1/\Delta MV_1)|_{both\ loops\ manual}}{(\Delta CV_1/\Delta MV_1)|_{loop\ 2\ in\ automatic}} \qquad (Eq.1)$$

A typical distillation tower has 5 CVs and 5 MVs; the system matrix is reduced to a 2x2 by assuming that pressures and levels are in control and that only the top and bottom composition, y and x respectively, are to be controlled by using 2 MVs. This simplification is possible based on the dynamic decoupling with the other CVs.

The procedure begins by having all the species in the feed distillate and bottoms assigned into a maximum of four components; two of them, Heavy Key (HK) in the Distillate, and Light Key (LK) in the Bottom product becoming the key components to be controlled. The other two are to group the lighter-than-light and heavier-than-heavy keys, LLK and HHK respectively. The theoretical number of trays (N) and the reflux ratio (RR=L/D) are the other two pieces of information required to calculate the relative gains that are presented in a tabular format to facilitate comparison among the possible pre-selected pairings. By combining material balance, Separation factor (S) and the Douglas et al [2] Separation model, the operating curves can be traced to demonstrate graphically how the product compositions change in the neighbourhood of the column products impurities specification, when the independent manipulated variable indicated is kept constant. All curve slopes around the impurity specification visually display what is calculated by the relative gains.

The Separation factor (S) – shown as Eq.2- for control purposes [7] is an extension of the Fenske-Underwood equation that calculates the minimum number of plates in a distillation column operating at total reflux. In this case, it is used to evaluate the split or, feed separation, into its components across the entire column as indicated by the ratio of the light key (y_{LK}) to the heavy key (y_{HK}) in the distillate (D) product composition specification, compared to the same quotient in the bottom (B) stream (x_{LK} and x_{HK}) when the column operates at other reflux conditions

$$S = \frac{y_{LK}/y_{HK}}{x_{LK}/x_{HK}} \qquad (Eq.2)$$

The Douglas Separation model shown in Eq.3, allows the calculation of RR once the compositions are known.

$$1/RR = \left[\left(\frac{\alpha}{S^{1/N}}\right)^2 - 1\right] \cdot z \quad (Eq. 3)$$

Where α is relative volatility, N the theoretical number of trays calculated by multiplying the actual number of trays in the column by the tray efficiency (N=n×E). Finally, z is the mol-fraction of the light keys in the feed. Efficiency of 70% has been the typical value used in the calculations.

Continuing with the case at hand, the species were distributed into four components. Keep in mind that the model requires that none of the LLK leaves through the bottom, that the LK's concentrate in the D, that the HK's concentrate is in the B, and that neither the HHK's leave through the top nor the LLK's through the bottom.

Accordingly:

- LLK is Toluene
- LK is the mix of Meta-xylene, Para-xylene and Ethyl-Benzene
- HK is Cumene
- HHK is C9+

In a variation in the use of the method, the ortho-xylene was excluded altogether from the distribution amongst the four components considered as it would appear in similar concentrations in the LK and HK.

The Feed, Bottoms and Distillate current compositions in %-mol, number of theoretical trays and RR complete the information entered as shown in the top table on the right hand side of Figure 1. The cells in green are entries and the yellow cells are calculated values. The table below that shows the calculated Relative Gains (RG) with the grey cell indicating that the impurities in the top product (y_{HK}) should be controlled by D/F, and that if the bottom composition was going to be controlled at the same time, then Boilup Ratio (VR) would be the first choice as the calculated RG, is closest to one. However, that would have required the tower bottom level to be controlled by manipulating the fuel gas to the fired heater. There are two quite important drawbacks to this: one, inverse response and secondly the limited operating range in the manipulated variable intended to control inventory.

The Feed, Bottoms and Distillate current compositions in %-mol, number of theoretical trays and RR complete the information entered as shown in the top table on the right hand side of the Figure 1; the cells in green are entries and the yellow are calculated values. The table underneath shows the calculated Relative Gains (RG) with the grey cell indicating that the impurities in the top product, y_{HK}, should be controlled by D/F, and that if the bottom composition was going to be controlled at the same time then Boilup Ratio (VR) would be the first choice as the calculated RG is the closest to one; that however would have required the tower bottom level control to be done by manipulating the fuel gas to the fired heater, against two quite important drawbacks: inverse response and limited operating range in the manipulated variable intended to control inventory.

The D/F ratio is 0.852, is in agreement with the intended operation of the tower to remove heavy keys from the petrochemical feedstock that exits from the top of the tower. The relative volatility does not have much practical meaning in this particular case, due to the multiple components of the feed and the exclusion of the ortho-xylene. This value can be used to run other scenarios such as the changes in product specifications and its implications in RR and energy requirements.

Figure 1 also shows that the DV strategy can be used to control the separation of the heavy components from the petrochemical feedstock. The nearly horizontal D/F means that by keeping this ratio constant, the top composition will stay very close to the specifications. The lack of specifications for the bottom product ruled out the otherwise severe interaction associated with dual-composition controls should the conventional LV pairing was used, as can be realized by the calculated RG of 156.

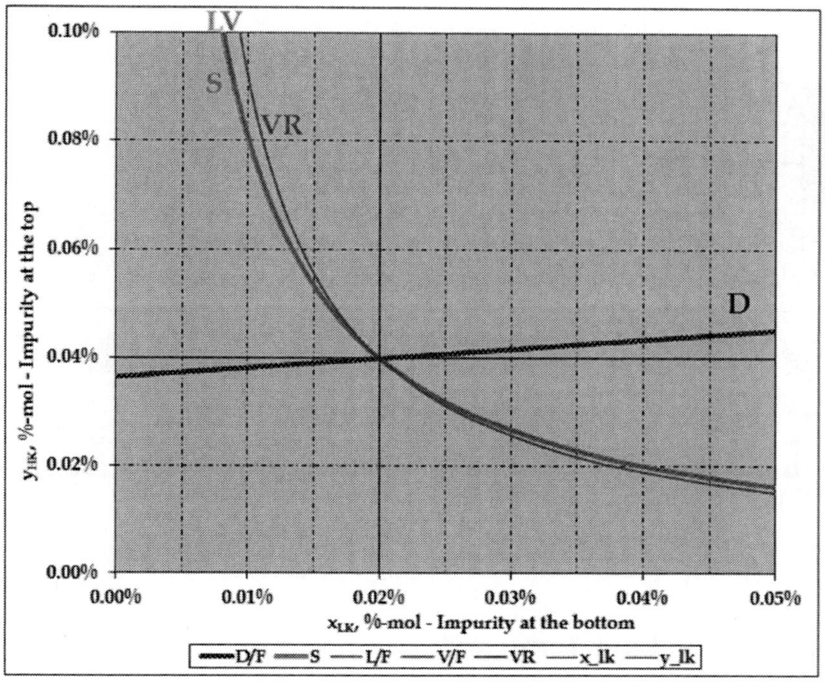

KEY		Feed	Distillate	Bottoms
		z%	y%	x%
Lighter	LLK	0.200	0.235	
Light	LK	85.000	99.725	0.020
Heavy	HK	14.000	0.040	94.563
Heavier	HHK	0.800		5.417
Theoretical Stages		100	D/F =	0.852
Reflux Ratio (RR)		1.5	α =	1.571

Relative Gains		MV2→y		
		D [D/F]	L [L/F]	S [L/D]
MV1→x	B [B/F]		0.079	0.080
	V [V/F]	0.921	156.011	62.957
	VR [V/B]	0.928	10.186	9.366

Figure 1 - Xylene tower - Specifications, RG and Operating Curves

The single-end composition control applied to the bottom stream released one degree of freedom and opened the door for what Shinskey [8] calls "local optimization" opportunities in distillation. When addressing this issue, he proposed an objective function that combines product losses and utility cost per unit feed. This shows that objective function minimization reduces the loss of the most valuable product and minimizes the heat input cost.

Extracted from the numerical application of this objective function as used in the Example 11 [Ref. 8, p.323], the relevant result for the purposes of this writing, has been plotted in Figure 2 to graphically demonstrate how the low Reflux/Feed ratio (L/F) sensitivity to feed composition changes can be used in the maximization of impurities in the top product. Explicitly: Once the optimum L/F ratio is found and maintained constant, it will keep the top composition nearly constant despite the feed composition variability. The arrows indicate the direction of different variables increase.

Because the numerical values of the example are not applicable here, the direction the different variables increase has been indicated by arrows.

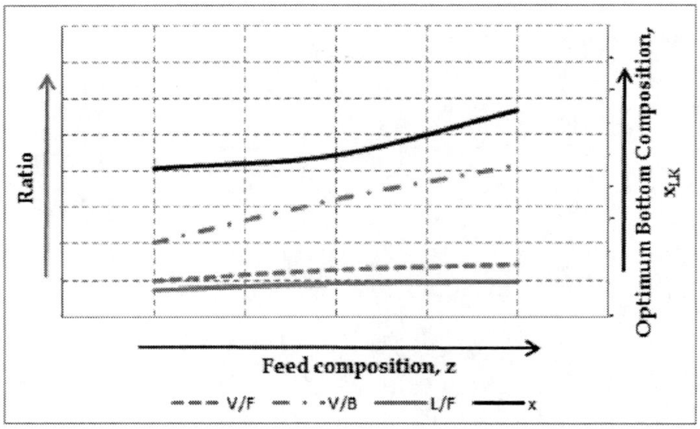

Figure 2 - Typical ratios Sensitivity to feed composition changes in a distillation column

The final control strategy, combining the top composition control and the optimization strategy is depicted in Figure 3. It includes two distinct dynamic compensation blocks: one in the direction of the top composition control and the second towards the reflux optimization loop. The high and low fuel gas valve openings are guaranteed by limiting the operating span of the reflux optimization controller (indicated by "Opt" in Figure 3), and a switch is available for the operator, should the fired heater operation controlling the coil outlet temperature is required, e.g. Cold Start-up. Although not shown in the drawing, there is a low oxygen control override to prevent the formation of fuel rich atmosphere in the fired heater; pass balance control was not required for the furnace in the application due to the inherent uniformity of temperatures and flows in the different passes.

Better control of the top of the column allowed the use of the optimization strategy to bring impurities from 0.04 up to 1.6, causing the D/F to increase. This additional volume produced at the top, multiplied by the difference in value between the petrochemical feedstock and the gasoline pool grade bottom product, accounts for about 70% of the total benefits. The remaining 30% of the profit is coming from the lesser V/F ratio resulting from the reduction the L/D ratio, directly reflected in fuel gas consumption reduction in fired heater reboiler. The total savings are in excess of 2 MM$/yr. for an ROI of 12 days. It can be reported that 5 years after implementation, the strategies are still in use, requiring only the routine field devices maintenance activities.

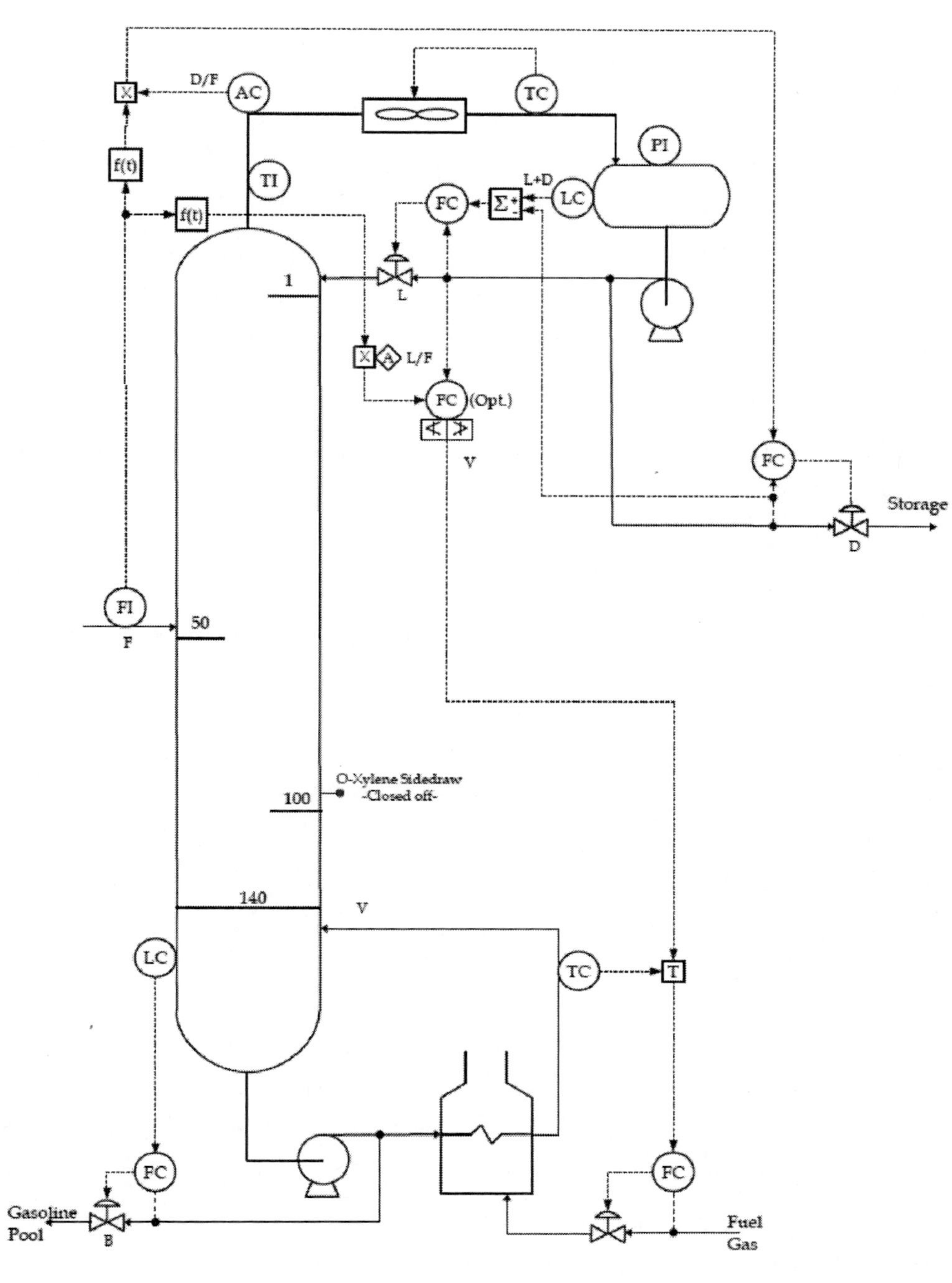

Figure 3 - Xylene tower - Control Strategies

CASE 2: Benzene-Toluene Heat Integrated columns – Why was ARC chosen over MPC?

A paper detailing the implementation of ARC-based dual-composition material balance control in a high-purity distillation heat integrated Benzene-Toluene tower application was published by this author in 2011[5]. The untold story, thus far, is that the ARC option was presented to the customer by a seasoned MPC engineer whose experience has been that in high purity distillation and other columns with long dynamics, it is quite difficult to implement model-based multivariable control and would still not perform as well as with ARC.

Dual-composition control in distillation columns has always been a challenging application. However, once the strategies are properly formulated based on the understanding of the interaction between material and energy balance in distillation, as well as the interaction associated with the CV-MV pairing, it is possible to select an array of control strategies to minimize interaction without sacrificing either robustness or performance of the operation.

There were multiple challenges in this application. Despite requiring a double-end composition, being heat integrated and having a ternary sidestream, amongst others, all difficulties were successfully overcome. After a recent visit to this installation, it can be reported that the application is still fully functional with the same control arrangement and tuning of the PID and dynamic compensators used in the commissioning over six years ago.

The pairing analysis done on the benzene tower by using Relative Gain (RG) showed the conventional LV - or energy balance to be the worst with an RG of 23.7 - 1.0 being optimal; compared to the material balance SV strategy that had a value of 0.998. The energy balance strategy instability driven by the LV based double-ended composition interaction was not affecting the benzene, but also the toluene tower through the heat integration.

One of the most counter-intuitive loops configured in the benzene tower was the control of the LK impurities in a tray below the feed tray. This configuration required the use of lead-lag compensation and the use of a considerable derivative time in order to improve the performance of the interacting-lag dominated dynamic response of the composition variable. This fact, along with the closure of the reboiler/condenser bypass valve highlights the importance of the thorough understanding of the process that must precede any control implementation.

As for the toluene tower, its RG analysis for the composition targets (Table 5, [5]) produced a 2.347 for the LV and 0.912 for the DV. Though it does not look like a sizeable difference, the biggest advantage of the material balance is that direct manipulation of the ratio D/F would result in a tighter and more robust option to minimizing the xylene losses. This advantage stems from the almost horizontal positioning of the material balance line showing that very tight control of the impurities at the top of the column is possible by keeping the D/F ratio constant. The inconveniently slow dynamic behaviour associated with the DV strategy is due to the lag imposed by the reflux drum. However, the use of the so-called Van Kampen decoupler [10] not only eliminates the problem but it can also convert the overhead accumulator into a lead, bringing the benefits of improved dynamics

Even though the original goal at the petrochemical facility was to bring stability to the operation of these towers that needed constant attention from the operators, the application ended up de-bottlenecking the operation and producing quantifiable benefits in excess of 1 MM$/year; for an ROI of 25 days. Also, it brought a noticeable increase in the much-sought stability of the operation.

CASE 3: Debutanizer – "The simulation shows that the conventional LV controls would work"

This tower was installed in an oil refinery. On one hand, its purpose is to control the volatility of the gasoline (Reid Vapour Pressure, RVP) during the warm months by limiting the amount of butanes in the bottom product. On the other hand, it had to ensure that the amount of pentanes in the distillate product were at a minimum.

The debutanizer was designed and commissioned for dual composition using the conventional energy balance configuration, LV. The decision to use LV as the control strategy was based on a computer simulation that determined this option would work. However, after several weeks of fruitless attempts to stop the continuous 1.5 hours period oscillation in both compositions, as depicted on the left-hand side of Figure 4, the refinery decided to implement a material balance control strategy.

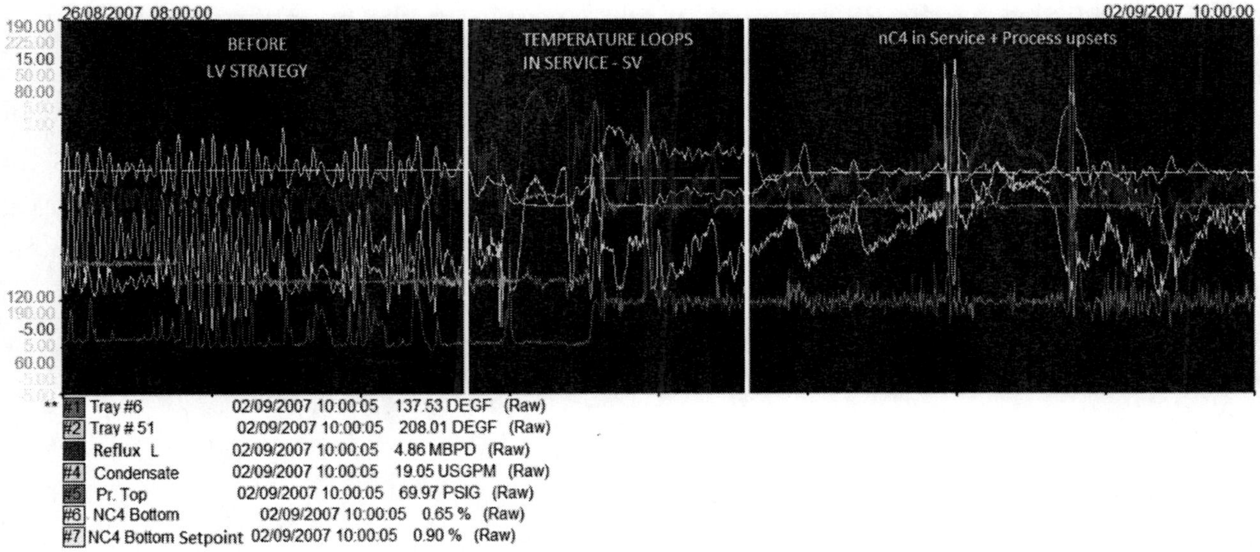

Figure 4 – Process Variables Trend – Before and After

As all the information required by the Shinskey's four-component model was available, it was possible to produce the relative gains and the corresponding operating curves. The following data was used for the specifications table: LLK is propane, LK is normal plus iso-butane, HK is normal plus iso-pentane, HHK is the balance.

The proximity of the L and V curves, shown in Figure 5, explains the strong interaction between the temperature control loops, confirmed by the value of 13.37 in the relative gain calculation. The chosen strategy is the SV as indicated by the grey cell in the RG table. The S-VR could have worked out as well; however, that would require controlling the tower bottom level with the heat input, which has the tendency to have an inverse response.

The control strategy, as implemented, can be seen in Figure 7, was proposed in 1981 by Ryskamp [6], and it features explicit and internal decoupling. The explicit decoupling between the reflux drum level control and the top composition control is performed by the summing block; when the top temperature controller changes the distillate flow (D), the reflux flow (L) will change by the same amount in the opposite direction. This way, the inventory remains constant. There is a partial internal decoupling between the controlled temperatures with the multiplier in the top temperature composition control circuit, as the RR stays constant when the controller output remains constant, thus preventing the top product composition from changing whenever there is a change in the heat input caused by the actions of the bottom temperature control.

Regardless of the PID optimal tuning parameters, the boundary will always be the proximity to the stability limit; once a controller has been brought up, the best practical performance limit can be pushed further by using feedforward control. Following the "adding only the required complexity provided that it outweighs the benefits" philosophy, the feed rate was added to the bottom temperature control loop as a feedforward through a dynamic compensator to pace the steam following a change in the feed.

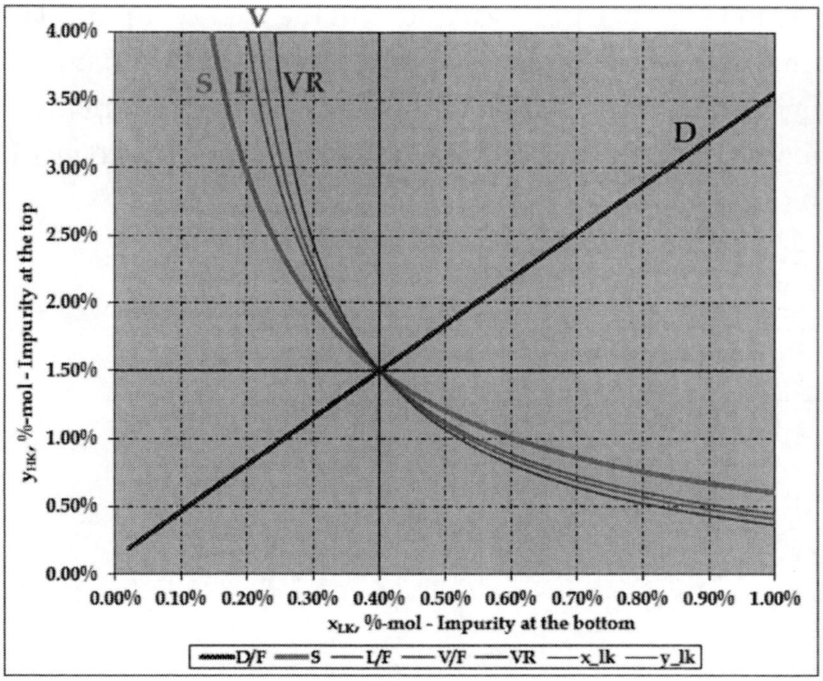

KEY		Feed	Distillate	Bottoms
		z%	y%	x%
Lighter	LLK	0.500	2.219	
Light	LK	22.000	96.281	0.400
Heavy	HK	50.000	1.500	64.103
Heavier	HHK	27.500		35.497
Theoretical Stages		35	D/F =	0.225
Reflux Ratio (RR)		3.7	α =	1.932

Relative Gains		MV2→y		
		D [D/F]	L [L/F]	S [L/D]
MV1→x	B [B/F]		0.412	0.481
	V [V/F]	0.607	13.371	3.323
	VR [V/B]	0.633	5.887	2.683

Figure 5 Debutanizer - Specifications, RG and Operating Curves

Although not explicitly illustrated in the column sketch, top and bottom controlled temperatures were Pressure Compensated by using the linear correction [8]:

$$T_c = T_{meas} - \left.\frac{\partial T}{\partial p}\right|_x \left(P_{meas} - P_{ref}\right) \quad (Eq. 4)$$

Where: T_c is the compensated temperature, T_{meas} is the measured temperature, P_{meas} is measured pressure, P_{ref} the reference pressure and the partial derivative is the slope of the vapour pressure curve as the normal product composition.

The tower propensity to flooding prompted the inclusion of a column differential pressure override (ΔPC-2 in the figure) in the strategy implemented in order to stop the heat input from increasing when the pressure drop across the column nears the incipient flood point.

The column came with three final control elements capable of controlling the tower pressure, which is not only an over-specification of the system but also the conflicting operation of variables that are not independent. This issue was highlighted at the time but was addressed only after the completion of this project.

Even if the end products are going to storage tanks, where a sustained oscillation in the product composition can average out, the cycling in the distillation column product compositions produces non-symmetrical oscillations in the energy input due to the nonlinear relation between energy and purity. Figure 6 depicts this situation as applied to this tower; assuming a uniform oscillation of 0.5% around the 1.0% impurity setpoint, it can be seen that much more energy is required to over-purify than to yield a lower column product concentration

.

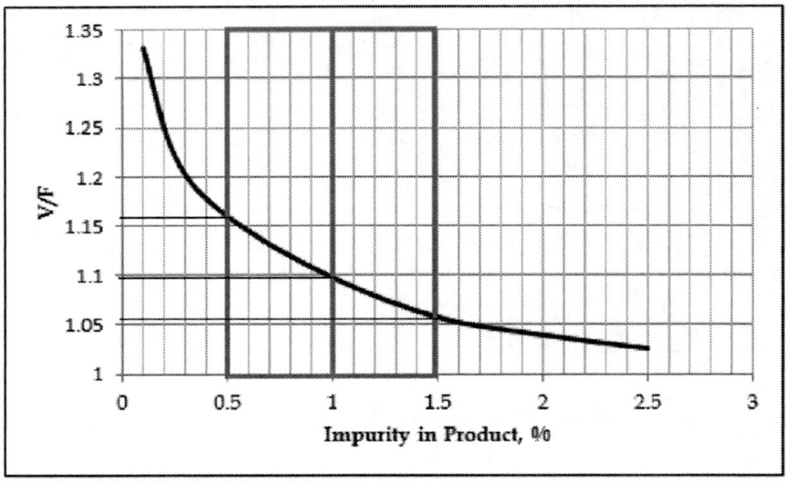

Figure 6 –Product Impurity VS. Energy Demand

The progressive improvement in the column captured in Figure 4 shows the transition toward the middle zone is probably the most impacting as the sustained oscillation in the column stopped immediately following the change of the top composition control strategy from reflux to reflux ratio. The right-hand side illustrates the steady state behaviour of the column operating under dual-composition control including instances of the ability of the control strategies to recover from upsets triggered by unmeasured disturbances.

As a result of this successful implementation of dual-composition controls, butane composition variability was reduced by almost 60%, making easier to maintain the RVP targets. The energy requirements were cut by 12%, or about $70000/yr. in a column that operates only 6 months per year. Lastly, because the tower was operated closer to the specifications, there has been a decrease of 14% in butanes lost to the bottom product.

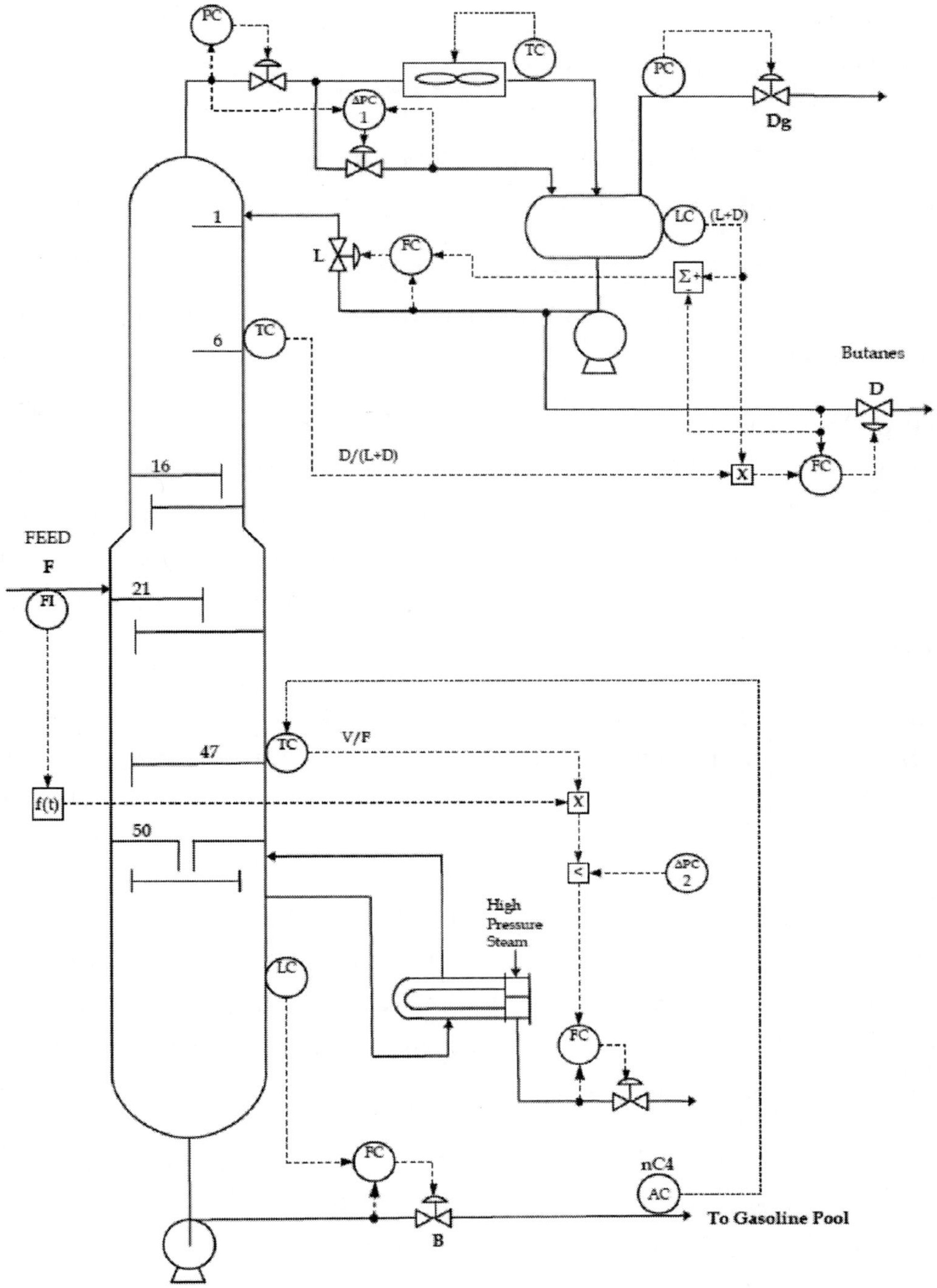

Figure 7 - Debutanizer - Control Strategies

FINAL REMARKS

Multiple applications of sound control practices applied to distillation columns, like those presented in this paper, have proven time after time that benefits can be accrued from improved stability and optimization.

Process control optimization can be the most lucrative investment, in terms of ROI, that can be done given that investment size is not as much as the level required for capital projects. An eye-opening Benefits/Investment ratio exceeding 10:1, achievable through process control engineering, should certainly be in the interest of any operation manager looking forward to maximizing the profitability of a process unit.

Optimization can be accomplished in single-composition controlled distillation column by minimizing the cost associated with its utilities and maximizing product recovery per unit feed.

Even if sometimes and oscillating PV can average out, the product composition-to-energy consumption nonlinear relation in a distillation column, causes a net excess in the heat input to the tower. Therefore, the elimination of oscillatory behaviour can result in immediate economic benefits because the cycling will cause the energy usage to be higher than it would be if the operation took place at the average values and compositions in steady state.

Distillation control is not beyond linear Advanced Regulatory Control (ARC). However, the added complexity in a control strategy is only justifiable if the benefits outweigh it. Simple controls work better and last forever - or until any of the field devices fail!

The estimation of the benefits of an optimized stable operation is difficult. However, in this author's experience, the optimization task is always profitable for the companies. Sound engineering based control applications have shown that once stability is attained, the opportunities for improvement will appear. In most cases, the benefits are well beyond expectations.

Once engaged in a plant project, the process control expert will have to find the way to work with the primary and final control elements that are already in place, setting aside the nice-to-haves. An initial survey can help, but unexpected situations will always arise.

MPC certainly has a place in process control; not as a rival to ARC but as a complement. The alleged ability of the MPC to work better than ARC, when there are complex dynamics involved, has always been baffling to the author; simply put, there are transfer functions that are uninvertible.

ACKNOWLEDGEMENT

The author wishes to dedicate this paper to F. G. Shinskey in recognition of his mentorship and support for over two decades, with the hope that his contribution to process control engineering will never be forgotten.

BIBLIOGRAPHY AND REFERENCES

[1] Bristol E. H. (1966) "On a New Measure of Interaction for Multivariable Process Control" IEEE Trans. Auto. Control, AC-11(1), p.133

[2] Douglas J. M., A. Jafarey, and T. J. McAvoy (1979):"Short-Cut Techniques for Distillation Column Design and Control. 1. Column Design", Ind. Eng. Chem. Process Des. Dev., V.18, No.2, pp.197-202

[3] Gordon, L. M. (1987) "Using Relative Gains to Design Dual Composition Controls". In Handbook of Process Control Systems and Instrumentation, Les Kane (Editor) Gulf Publishing

[4] McMillan, G. and S. Weiner (2015) "Is PID the algorithm of the future?" Control, June 2015, pp. 55-56

[5] Nino, S. (2011) "Improve material balance in high-purity distillation control". HPI, November

[6] Ryskamp, C.J. (1980) "New strategy improves dual composition column control". HPI, June.

[7] Shinskey, F.G. (1977) "Distillation Control" Second Edition, McGraw-Hill, New York, pp. 41

[8] Shinskey, F.G. (1984) "Distillation Control: For Productivity and Energy Conservation" Second Edition, McGraw-Hill, New York

[9] Shinskey, F.G. (2009) "Multivariable Control of Distillation", Parts 1, 2 and 3. Control Global, May, June and July

[10] Shinskey, F. G. (2012) "Control Systems: Investment and Return". Expertune website

[11] Van Kampen J.A., (1965) "Automatic Control by Chromatographs of the Product Quality of a Distillation Column". Proc. Instn. Mech. Engrs. 1964-65, Vol. 179 Pt. 3H

481695 Details Matter When Designing Hot Vapor Bypasses

Abstract:

Valero's Meraux refinery has a Hydrocracker that is a world-class high conversion unit which converts heavy gas oil (HVGO) and Deasphalted oil (DAO) into higher value products; primarily ultra-low sulfur diesel (ULSD).

In 2014, the unit was upgraded to improve operating margins with changes to the unit design. Modifications were completed in the feed system, a reactor was added, and major changes in the fractionation system. Fractionator changes included changes to improve capacity and fractionation efficiency to handle the new product slates. As part of the revamp, additional heat removal was added to the condenser circuit that increased the capacity by over 50%.

After startup, it was apparent that the hot vapor bypass system was not operating as designed. Primary fractionator pressure control was unstable, along with major process fluctuations in circuit temperatures and drum level. The instability was severe enough that the refinery considered a shutdown of the newly re-started hydrocracker to avoid potential equipment damage.

Field troubleshooting identified the problems with the overhead drum design which resulted in hot vapor bypass instabilities with a highly subcooled system. After identification, the overhead trim condenser duty was reduced by reducing water flow to limit subcooling. This eliminated the instability until a pad gas system was installed.

The author will present operating data that substantiates a previously unseen phenomena with hot vapor bypass pressure control configurations. Design parameters and recommendations will be made to improve future hot vapor bypass system designs.

Silencing Hammering in a Flooded Condenser System

By

Henry Z. Kister, Fluor, Aliso Viejo, CA, USA

and

Casey Mueller and Matthew Gunn, HollyFrontier Refinery, Cheyenne, Wyoming

Presented at the Kister Distillation Symposium, AIChE Spring Meeting, San Antonio, Texas, March, 2017

SILENCING HAMMERING IN A FLOODED CONDENSER SYSTEM

Henry Kister, Fluor, Aliso Viejo, CA, Casey Mueller and Matthew Gunn, HollyFrontier, Cheyenne, Wyoming

SUMMARY

Flooded condensers are, and are likely to remain, the prime tower pressure control methods for total condensers generating liquid product only. The principles of these methods had been described in the literature more than 60 years ago, yet these methods continue to rank among the most troublesome distillation controls. Good understanding of the principles and learning from experience is the key for avoiding potential traps and improving operation for the industry. This paper describes a recent experience solving a problem that, until now, has been unknown and poorly understood in the industry.

In its first year in operation, a naphtha splitter experienced intermittent hammering in the line from the tower overhead flooded condenser to the reflux drum. The problem occurred at cold condenser outlet temperatures. In some of the hammering episodes, the hammering was severe, opening a flange and causing damage to a valve and a thermocouple.

Fluor, which was not involved in the original design, was requested to join the HollyFrontier task force investigating the hammering incidents. Our task force identified the large open slots at the top of the reflux drum dip pipe as a likely root cause. We also identified total closure of the tower overhead pressure control valve as another potential source of hammering.

Based on this diagnosis, we blanked the slots near the top of the dip pipe and added a DCS clamp on the overhead pressure control valve that prevents the valve from closing to less than 20%. HollyFrontier also emphasized in its operator training the importance of keeping tight control of the condenser outlet temperature at 180° F. The column was returned to service in March 2015, with no further hammering to date, and experiencing smoother operation.

This paper describes our investigation and our solution.

BACKGROUND

The HollyFrontier naphtha splitter had been installed to help the Cheyenne Refinery comply with the requirements of the MSATII rule for benzene in gasoline by removing benzene precursors in the feed to the naphtha reformer (a UOP Platformer ®). The tower was started up in February, 2014.

In its first year of operation, intermittent hammering occurred in the line from the tower overhead condenser to the reflux drum. The problem occurred at cold condenser outlet temperatures. In some of the hammering episodes, the hammering was severe, opening a flange and causing damage to a valve and a thermocouple. During the winter of 2014/2015, the naphtha splitter was shut down and HollyFrontier requested Fluor to join its task force to investigate the hammering incidents with the objective of implementing a solution prior to restart.

PROCESS DESCRIPTION

Figure 1 shows the HollyFrontier naphtha splitter. Process conditions shown are those at 10:30 pm on November 10th, 2014, just prior to the incident described below. Feed to the tower is the naphtha hydrotreater (NHT) debutanizer bottoms. This feed is preheated to 290° F by exchange with the tower bottoms. The preheated feed then enters the 9' ID, 70 tray naphtha splitter. The tower separates a light naphtha overhead product from a heavy naphtha bottom product. The tower is reboiled by a fired heater. Tower bottoms are pumped through the feed preheaters, an air cooler and a sulfur sorber to the Platformer. Tower overhead vapor at 217° F is condensed by an air condenser, which is a total condenser (no vapor product), and the condensate flows into the reflux drum. From the drum, most of the condensate is returned to the tower as reflux, and the rest is pumped to storage via an air cooler.

Tower pressure is controlled at about 29 pounds per square inch gauge (psig) using a control valve in the tower overhead line to the air condenser. There is a bypass around the condenser with a control valve that is manipulated by the differential pressure between the tower and the reflux drum, kept at 5-10 psi.

The condensate line leaving the condenser enters at the top of the reflux drum and is extended inside the drum by a dip pipe that takes the condensate to near the bottom of the drum. To raise tower pressure, the pressure control valve in the vapor line to the condenser is throttled. This reduces the pressure in the condenser, which in turn reduces the delta T (temperature difference between the condensing side and the cooling side) and also sucks liquid back from the drum. This liquid floods some of the condenser tubes, which reduces its condensation area, in turn reducing condensation, causing the tower pressure to rise. Conversely, opening the pressure valve lowers the condensate level, exposing more condenser tube area for condensation, and also raises the condenser pressure, which improves its delta T. Both increase the condensation rate and lower tower pressure.

The condenser bypass line helps keep up the pressure in the reflux drum. If the valve in the bypass line is closed, the pressure in the drum will be the vapor pressure of the condensate liquid. When the condensate is near its boiling point (no subcooling), the drum pressure will be much the same pressure as the pressure at the condenser outlet. In contrast, when the condensate is subcooled, its vapor pressure may be much lower, typically by about 0.4 psi per degree F of subcooling in naphtha splitters. The hot vapor bypass (HVB) line and valve prevent the drum pressure from falling too low. When condensate subcooling causes the pressure difference between the tower and the drum to rise, the HVB valve opens and diverts additional hot vapor to the drum. The hot vapor warms up the drum liquid surface, making the drum pressure equal to the vapor pressure of this warmed-up liquid surface (which is warmer than the subcooled bulk liquid). Process liquids are good thermal insulators, so, as long as the subcooled liquid does not reach the surface in excessive or fluctuating quantities, the surface temperature will remain steady and warm enough to sustain the differential pressure set by the dP controller. A detailed discussion is included in Reference 1.

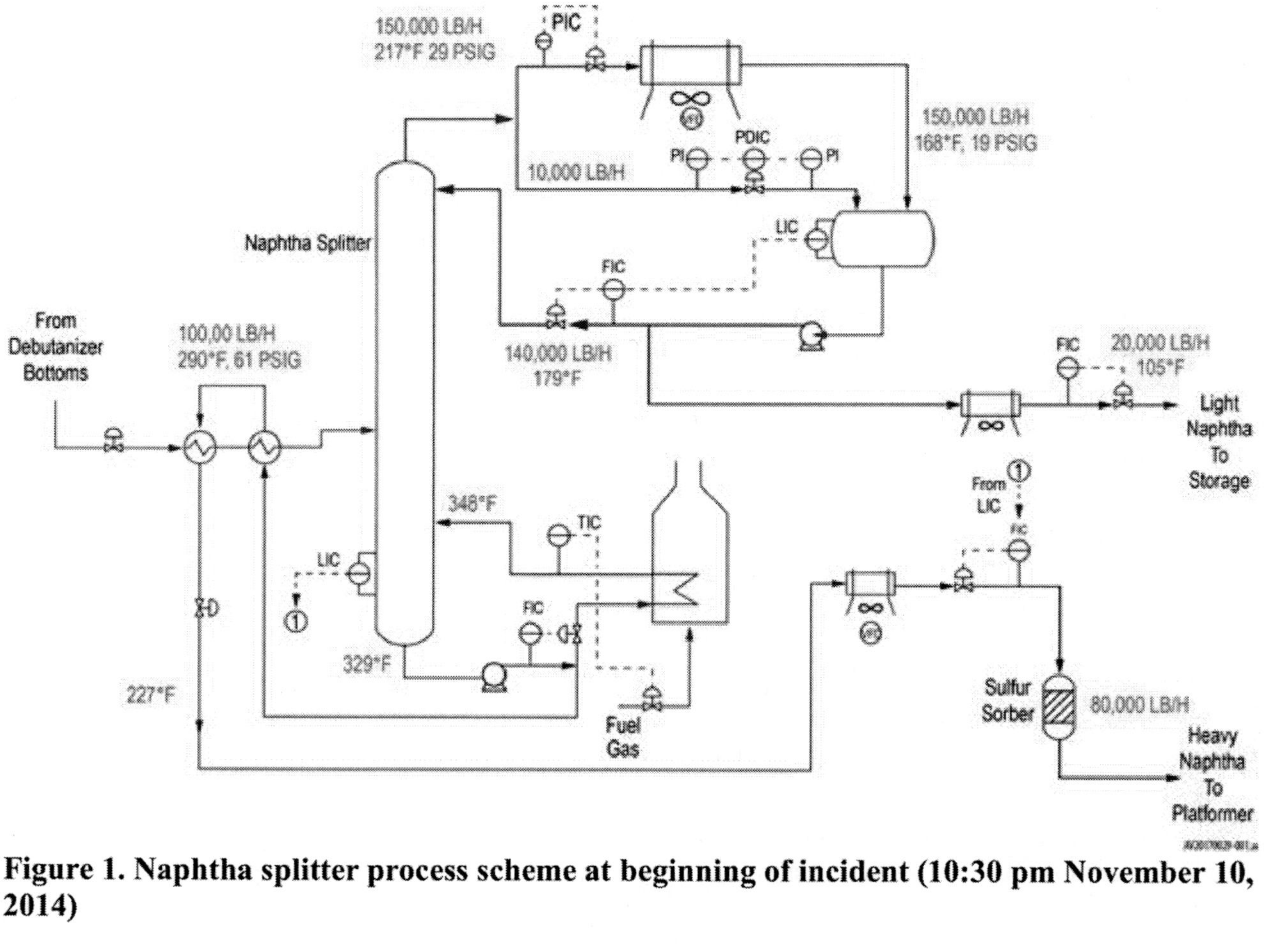

Figure 1. Naphtha splitter process scheme at beginning of incident (10:30 pm November 10, 2014)

FOCUS ON THE CONDENSER AND REFLUX DRUM

Because the hammering occurred between the condenser and the reflux drum, this area was the focus of our investigation. Figure 2 shows the physical layout of the piping from the air condensers to the reflux drum. Both the vapor and condensate lines descend to the drum with no pockets according to good design practices (1).

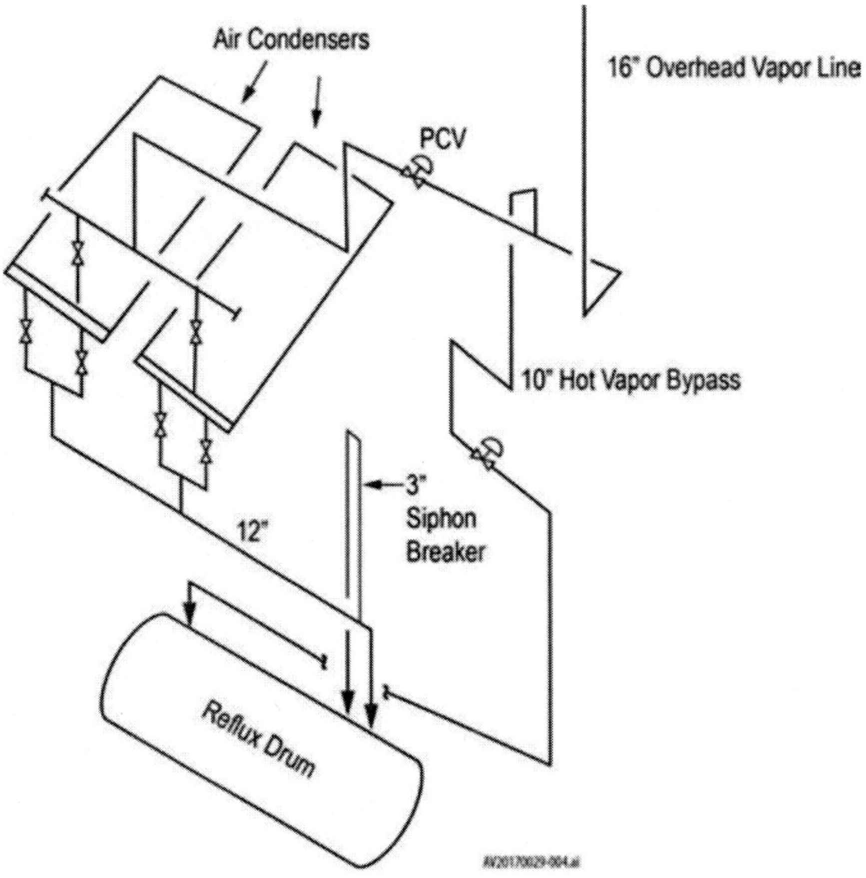

Figure 2. Layout of the naphtha splitter condenser and reflux drum piping

Figure 3 shows the reflux drum. The drum is 6.5 ft ID by 20 ft tangent to tangent horizontal drum. Condensate enters the drum via the 12 inch nozzle on the top right hand side of the drum in Figure 3. This nozzle is equipped with a dip pipe shown in detail in Figure 4. Condensate leaves the drum via the 8" nozzle at the bottom left of the drum in Figure 3. The vapor bypass around the drum enters via a 10" nozzle at the top left of the drum in Figure 3.

Figure 4 is a close up of the dip pipe. The pipe terminates 1 foot above the bottom of the drum. The bottom of the dip pipe is open. The dip pipe is equipped with four 2" wide by 12" long slots, starting 6" below the top of the drum, and continuing 12" to 18" below the top of the drum.

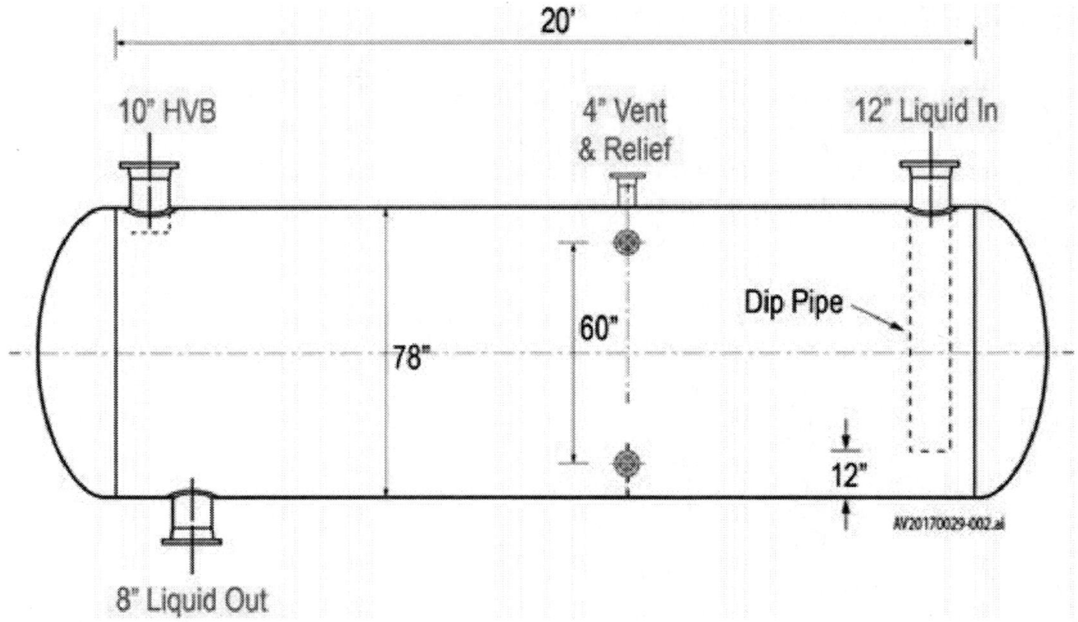

Figure 3. Naphtha splitter reflux drum

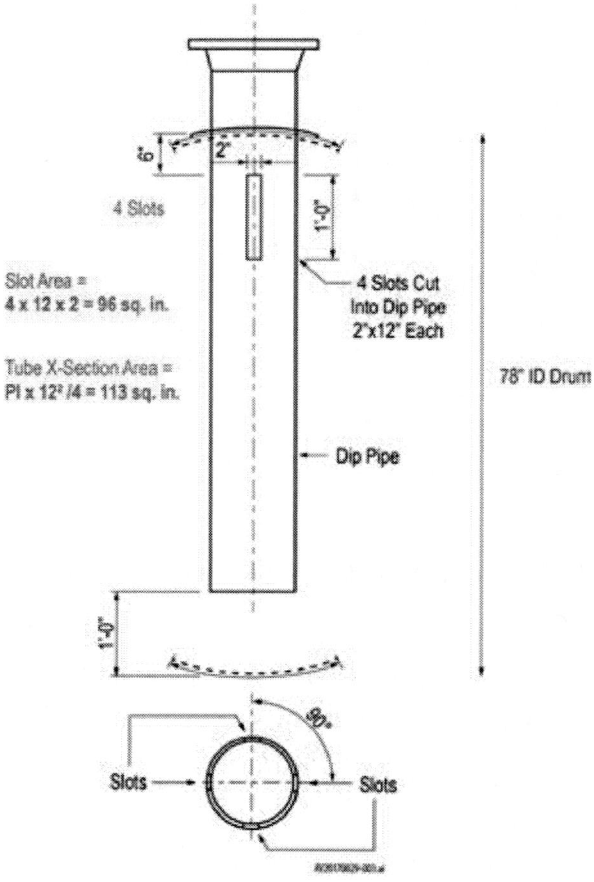

Figure 4. Naphtha splitter reflux drum dip pipe

During the design of the Naphtha Splitter, a concern was raised that non-condensables can be present in the column feed at times. The upstream Debutanizer may occasionally lose reboil, resulting in light ends (butanes and lighter) being present in the feed to the Naphtha Splitter. During such events it was envisioned that these light ends would need to be vented from the overhead accumulator; accordingly, slots were added to the dip tube to allow vapors to disengage from the liquid at the outlet of the condenser.

Emphasized in Figure 4 are the areas of the slots compared to the area of the opening at the bottom of the dip pipe. The slot area is 96 sq. inches, while that of the opening at the bottom of the pipe is only slightly higher, 113 sq. inches. The middle of the slots is 4.5 ft above the bottom of the pipe. When the drum level is about 50%, the additional head acting on the bottom of the pipe is about 2.25 ft higher than that acting on the slots. The total liquid head acting on the slots is therefore about 16 ft, compared to a head of about 18 ft acting on the bottom of the pipe. With the liquid heads being close, the liquid split between the slots and the pipe bottom will roughly follow the opening areas, 46% through the slots, 54% through the bottom of the pipe. That is without considering any vapor pressure effects.

OPERATING EXPERIENCE

The naphtha splitter was started up in February, 2014. It has been operating at rates close to those shown in Figure 1. These rates are below the future design rates of the system.

In its first year in operation, intermittent hammering occurred in the line from the overhead condenser to the reflux drum. The problem was first observed following the February 2014 startup, and occurred at cold condenser outlet temperatures. No hammering was observed when the condenser outlet temperature exceeded 170° F. The hammering was alleviated by tuning the condenser outlet temperature controller (which adjusts the speed of the fans motors) to avoid rapid changes in condenser outlet temperature.

There were two incidents where the hammering was particularly severe. During an outage in the summer of 2014, an effort was made to operate the tower at total reflux, however, some light naphtha product was still drawn. This resulted in heavier tower naphtha with overhead temperature reaching 300° F (normally 217° F). When the reboiler was shut down, the tower pressure appeared to collapse. The result was such strong hammering in the overhead line that flanges opened and the handle of one of the 6" valves at the condenser outlet came off.

A second strong hammering incident occurred during an outage on a cold night on November 10th, 2014. Again, the hammering followed the shutdown of the reboiler. In this incident, the hammering damaged the thermocouple used to control the air condenser fans. A decision was made to not restart the column until a solution to the hammering problem could be implemented.

In addition to these two strong hammering incidents, a number of several smaller hammering episodes occurred prior to the November, 2014 shutdown. In these incidents, the plant experienced condenser outlet temperature swings of 40-50° F, sometime falling from 170° F to

100° F in two minutes. Drum liquid temperature would drop by up to 20° F. At times the line from the condenser to the drum heated up.

Even at periods of no hammering, the tower overhead system appeared to experience frequent instability.

Following the November, 2014 naphtha splitter shutdown, HollyFrontier requested Fluor, who was not involved in the initial design but has wide distillation expertise, to join its task force investigating the hammering incidents, with the objective of implementing a solution prior to restart. Based on a review of the system, an event timing analysis, and conducting a comparative analysis using previous experiences with which Fluor has familiarity, the taskforce identified the problem and implemented a solution.

EVENT TIMING ANALYSIS

The event timing analysis was applied to an incident in which tower feed was lost between 10:30 pm and 11:20 pm on November 10, 2014. During this period, the tower feed rate dropped from 100,000 lb/h at 10:30 pm to 42,000 lb/h at 10:45 pm, reaching zero at 11:10 pm. Hammering in the aftermath of this incident later that night became so severe that it broke the condenser outlet temperature indicator.

The key parameter emerging from the event timing analysis was the tower overhead flow rate to the condenser, which was calculated from the pressure control valve opening. The variation of this flow rate is represented in Figure 5.

Between 10:30 pm and 10:45 pm, while the feed rate dropped from 100,000 to 42,000 lb/h, the bottom level held constant at 55%, and the bottom flow rate dropped from 80,000 lb/h to 61,000 lb/h. The reboiler duty was constant with the fuel gas rate to the reboiler at 40 MCFH.

The much larger reduction in feed rate compared to the bottom flow rate led to increased heating of the feed in the preheater, with the preheater outlet temperature rising from 290 F°to 304° F. This led to more vaporization in the feed and more tower overhead vapor. Based on the overhead control valve opening, the tower overhead flow rate rose from 150,000 lb/h at 10:30 pm to a peak of 157,000 lb/h at 10:45 pm.

Over the next 10 minutes the fuel gas flow to the reboiler dropped from 40 MSCFH to zero, leading to a decline of tower temperatures. The tower overhead vapor flowrate plunged. Based on the overhead control valve opening, the tower overhead flow rate declined to 123,000 lb/h by 11:00 pm and to 98,000 lb/h by 11:10 pm. Boil-up stopped and the bottom and distillate flow rates dropped to near zero. Instability set in about 11:10 pm. Tray temperatures dived, the condenser outlet temperature swung approximately 30° F peak to valley every 2-3 minutes, and with it the differential pressure between the tower overhead and the reflux drum swung 4-8 psi peak to valley. By 11:20 pm the tower overhead flow rate declined to 62,000 lb/h with the swings continuing.

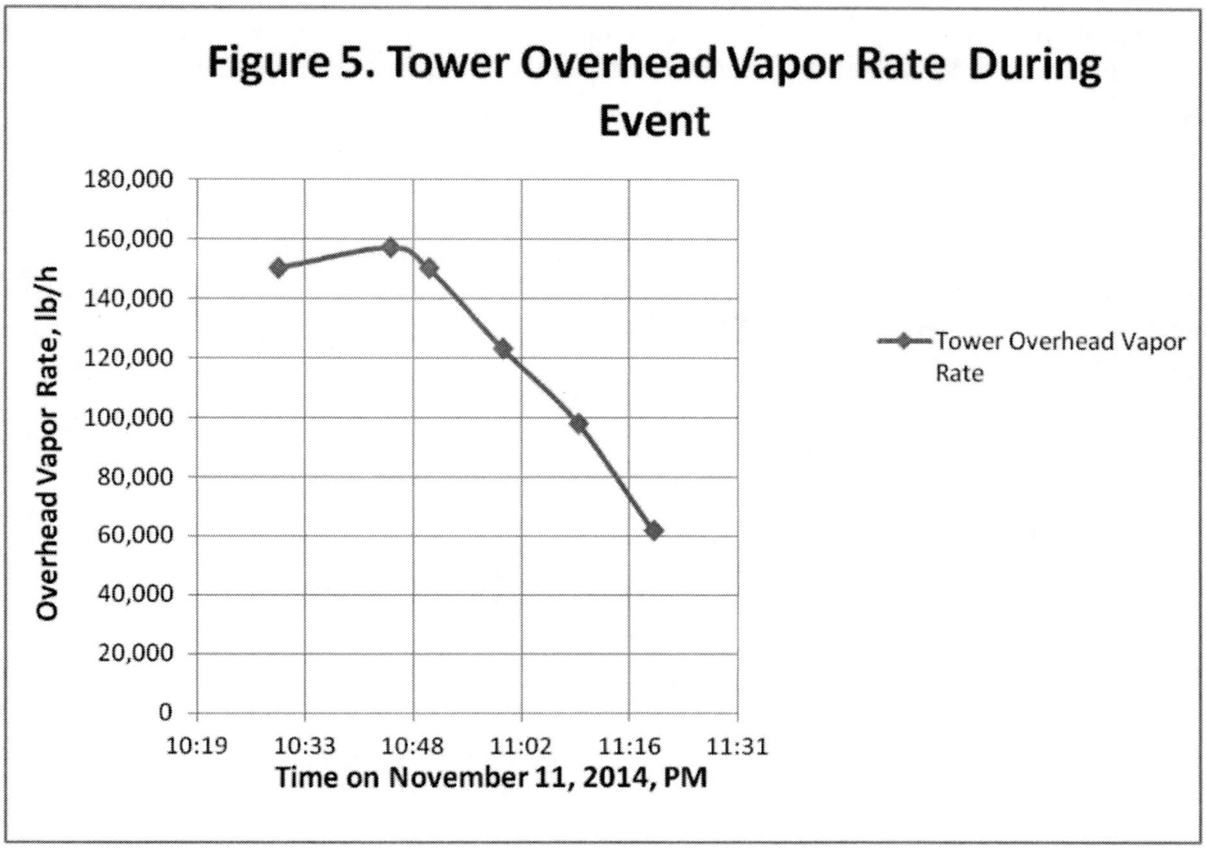

CONDENSER PERFORMANCE DURING THE EVENT

The overhead air condenser has two bays, each bay containing 238 34' long 1" OD finned 2-pass tubes arranged in four rows. The total bare tube area is 4240 ft^2. The tubes are sloped at 0.125" per foot, in accordance with good design practice. There are two fans per bay. The condenser outlet temperature controller manipulates the speed of the motor of one fan per bay. The other fan for each bay is on an on/off control mode. There are also louvers than can be manually adjusted to help maintain the desired condenser outlet temperature.

The condenser design case shown on the rating sheet has the splitter overhead temperature at 192° F, about 25° F colder than the splitter overhead temperature of about 217° F at the beginning of the November 10, 2014 event. Assuming the tower overhead is fully condensed at about 180° F (approximately the reflux temperature at the beginning of the November 10, 2014 event), the design case utilizes about 93% of the condenser area for condensing, and the remaining 7% for sub-cooling. The log mean temperature difference (LMTD) for the condenser design rating sheet was 55 F.

The condenser design rating sheet is based on a summer air inlet temperature of 95° F. On November 10th the ambient air temperature in Cheyenne was 10° F. With air temperature this cold, two of the four fans would normally be turned off, the others variable speed drives used for control. So the air flow would be approximately half the design, possibly even less. Even with two fans off, we estimated the LMTD to be about twice the design, around 100° F. This means that the condenser area utilized for condensation would be half or less the available condenser tubes area. Half the tubes would then be flooded. In the flooded area, the high delta T will give tremendous subcool.

Figure 6 shows our estimate of the number of unflooded tubes (i.e., those that perform condensation) during the event. When this number is 4, the condenser area is fully utilized for condensing, and the subcooling area is minimal. Figure 6 shows that at the beginning of the November 10 event (10:30 pm), until about 10:50 pm, there was little change in the number of unflooded tubes. After this, however, there was a rapid decline, with one row of tubes becoming submerged in liquid every 20 minutes or so. By 11:20 pm, the number of unflooded rows of tubes would decline to approximately 0.7.

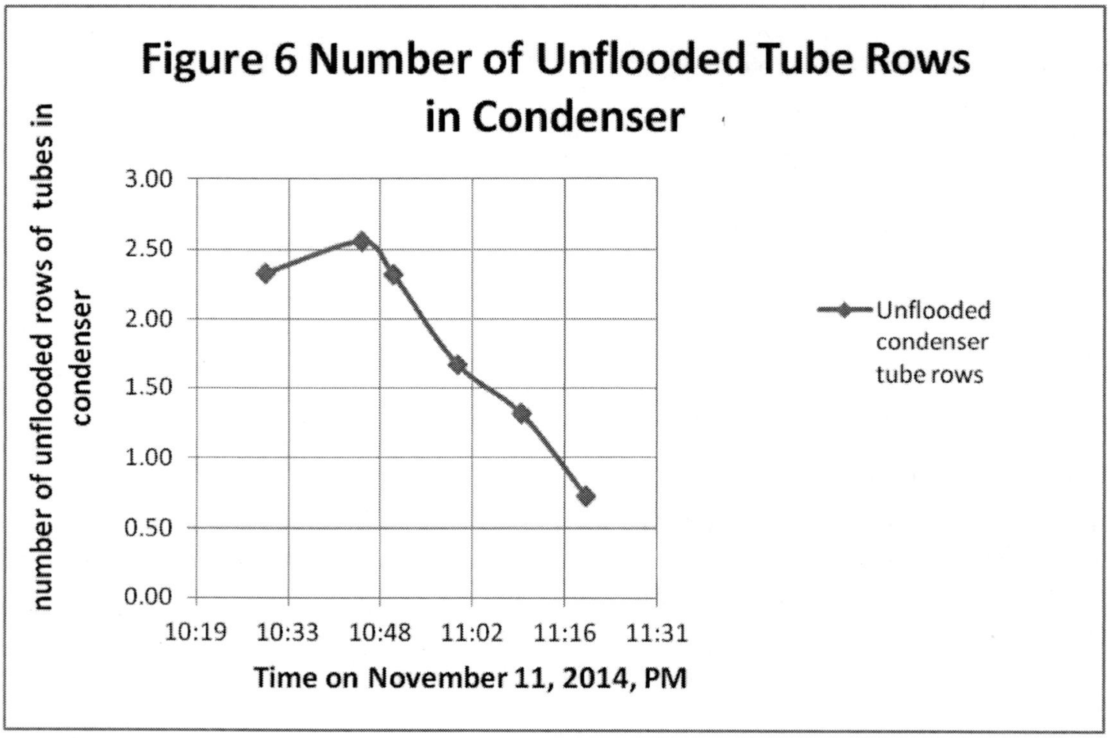

The scenario described in Figure 6 and in the two preceding paragraphs above is not realistic. It is based on an assumption that, on a cold night like November 10th, about half the normal air flow ascends through the air cooler. In reality, however, the condenser outlet temperature control and the operating crew will keep reducing the speed of the two operating air condenser fans to keep the subcool (as measured by the condenser outlet temperature) the same. Because subcooling increases as tubes flood, the control and operators will act to keep the liquid level at the air cooler constant. This also means keeping the number of unflooded tubes close to 4. It does not take much to subcool the liquid. We estimate that flooding less than 0.1 row of tubes will give a subcool of 10° F, even at the design (summer) conditions. With the cold air on the night of November 10th, flooding less than this will easily give 10° F subcool.

The only levers that the operators can pull are the fan speeds and louver openings. As the air gets colder and the duty declines, they would reduce fan speeds. The plant seldom manipulates the louvers to reduce air flow (rightly so) as this is an awkward way of controlling air condensers and can cause damage to the louvers. However, even if the operators turn off all fans, and even if they close all louvers, there would be a draft coming through the air condensers. If the louvers were left open, this draft will be quite strong. Figure 7 shows that before 11 pm on November 11th, there would have been quite a good condenser duty, and normal fan manipulation would have kept the subcool under control with condenser outlet temperature of about 170° F. At about 11:10 pm, the condenser duty dropped to just over half of the duty at 10:30 pm (Figure 7), and kept dropping as the reboil was stopped. With that rapid a decline in reboiler duty, it became difficult, if not impossible, to contain the subcooling, and the condenser outlet temperature dove precipitously as can be seen in Figure 8.

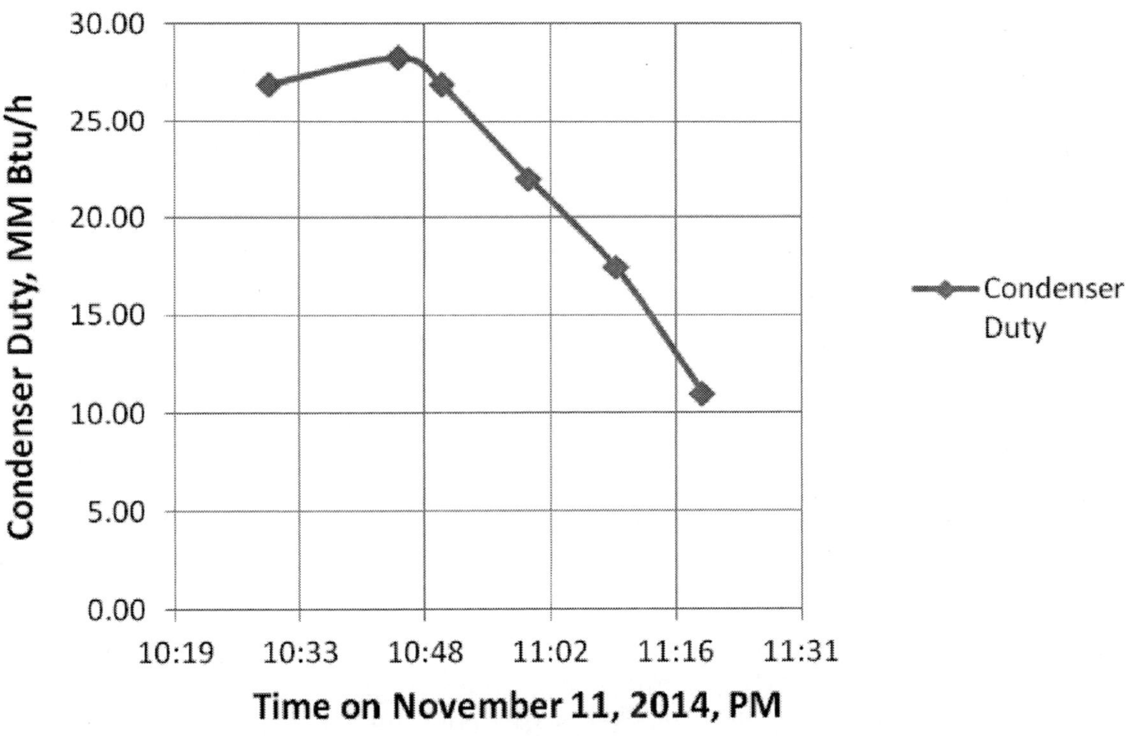

Figure 7 Condenser Duty During Event

The liquid head between the condenser level and the middle of the slots in the reflux drum dip pipe is approximately 16 ft, equivalent to 4.2 psi. The naphtha vapor pressure changes about 0.4 psi per degree F. Accordingly, as long as the subcooling is less than about 11° F (= 4.2/0.4), the liquid head will exceed the difference in vapor pressures and there will be flow from the tube out. At larger subcool, the difference between the drum pressure and the vapor pressure of the liquid in the dip pipe will exceed the liquid head, generating a potential for vapor backflow from the drum into the dip pipe. Initially, the momentum of liquid exiting the slots as well as suction of some liquid from the bottom of the drum to the slots will prevent the back flow. However, as subcooling intensifies, the driving force (pressure difference between the drum pressure and the vapor pressure of the subcooled liquid) will increase. Eventually, a point will be reached when the driving force will be large enough to overcome the liquid head, and induce backflow of vapor from the drum into the dip pipe. This dew point vapor will collapse onto the subcooled liquid in the condensate pipe. When this vapor collapse occurs in the confines of the pipe, hammering will result. The vapor backflow may heat up the condensate pipe, as was observed during some of the hammering incidents.

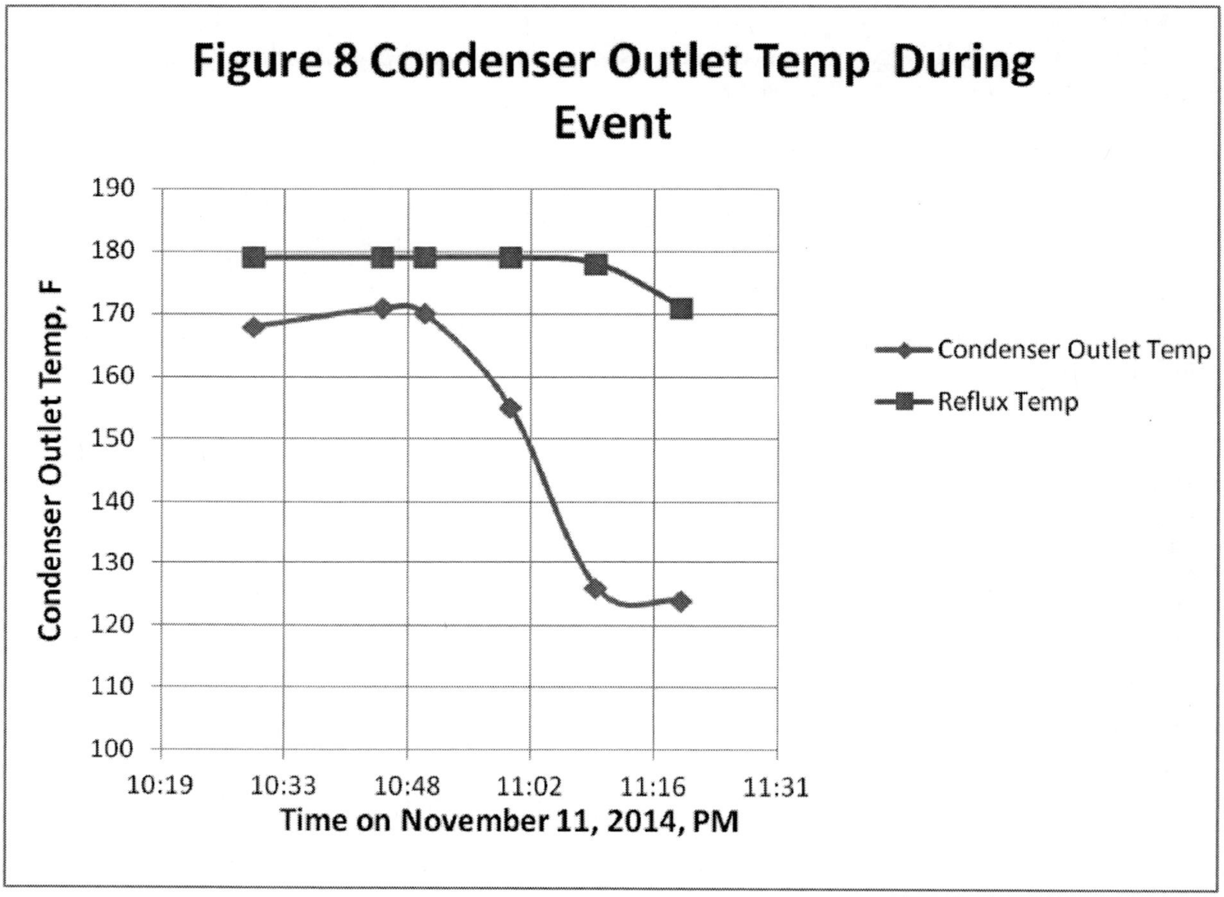

Figure 8 shows that until about 10:50 pm, there was a constant but low degree of subcool. Assuming the tower overhead is fully condensed at about 180° F (approximately the reflux temperature at the beginning of the November 10, 2014 event), this subcool was about 10° F. The hydraulic head exceeded the vapor pressure difference, so flow was from the slots in the dip pipe into the drum. Between 10:50 pm and 11:00 pm, there was a 15° F drop in the condenser outlet temperature, raising the subcool to 25° F. At this subcool, there was a 10 psi pressure difference between the drum pressure and the vapor pressure of the subcooled liquid, exceeding the head difference. However, from the plant charts it appears that this was not enough to initiate backflow of vapor from the drum to the dip pipe, as no hammering occurred.

From 11:00 pm to 11:10 pm, the condenser outlet temperature dropped by another 29° F, generating a subcooled liquid with a vapor pressure 22 psi less than the drum pressure. The drum pressure at 11:00 pm was steady at about 17° psig, in equilibrium with the liquid in the drum. By 11:10 pm, the vapor pressure of liquid in the pipe dropped to about 10 psia. This cold liquid exerted a strong suction on the drum vapor, much higher than the liquid head in the pipe. Therefore, instead of the liquid from the pipe pouring into the drum vapor space, this huge suction sucked vapor into the pipe. The dew point vapor from the drum rapidly collapsed onto the subcooled liquid, causing the hammering.

Had the slot area of the dip pipe been small compared to the open area at the bottom of the pipe, the resulting subcooled liquid that was sucked into the slots would have been less significant, resulting in minimal or no hammering. However, here the slot area was large, about 85% of the tube bottom area, so with high subcool, the suction could not match the vapor pressure difference.

Based on the above analysis, the key to prevention of further hammering incidents is blanking the slots in the dip pipe and/or limiting the subcool. Limiting the subcool can be readily achieved during normal operation, but may not be achievable during startups/shutdowns.

PRESSURE CONTROL VALVE CLOSURE

Complete closure of the tower overhead pressure control valve has been responsible for a number of severe hammering incidents such as Case 28.6 in Reference 2. If this valve fully closes, the cold air (especially during low rate operation and/or winter time) will quickly condense the small amount of vapor present downstream of the valve. The condenser pressure will immediately drop, and may go into vacuum, causing liquid to be rapidly sucked from the accumulator drum. The rushing liquid will bang against any objects in its path, producing a severe liquid hammer.

Figure 9 is a plot of the naphtha splitter tower overhead pressure controller output throughout the event. The diagram shows that at the lowest signal, the valve output was 38%, which coincides with an overhead vapor flow rate of 62,000 lb/h. A review of the operating charts during the hammering incident, throughout the night (until 4:30 am on November 11$^{th)}$ shows that the smallest valve opening recorded during this event was again 35%. We therefore conclude that during the analyzed event and the severe hammering that followed and damaged the thermocouple, the control valve did not get anywhere near total closure, so this mechanism did not contribute to the observed hammering on November 10th and 11th.

The observation that the overhead valve was kept away from closure probably resulted from good operation of the overhead condenser, where operators switch off fans and minimize fan speed during the low rates and cold weather. However, although valve closure did not occur in these cases, it does not preclude the possibility of an event in which the overhead valve will attempt to close fully during cold weather. We therefore added a DCS clamp that will prohibit the valve from reaching full closure.

POST-REVAMP OPERATION

Based on our joint task force review, we blanked the slots near the top of the dip pipe and added a DCS clamp on the overhead pressure control valve that does not allow the valve to close to less than 20%. HollyFrontier also emphasized in its operator training the importance of keeping tight control of the condenser outlet temperature at 180° F.

The column was returned to service on March 12, 2015 and operated continuously until May 13, when it was again shut down due to a refinery maintenance outage. It has been restarted and shut down on a number of occasions since then. No further hammering has occurred during normal operation nor during any of the startup/shutdown events, even though condenser outlet temperatures were as low as 100° F, and even less during startups and shutdowns. During normal operation, the condenser outlet temperature is always kept close to 180° F.

Figure 9. Condenser Valve Output During Event

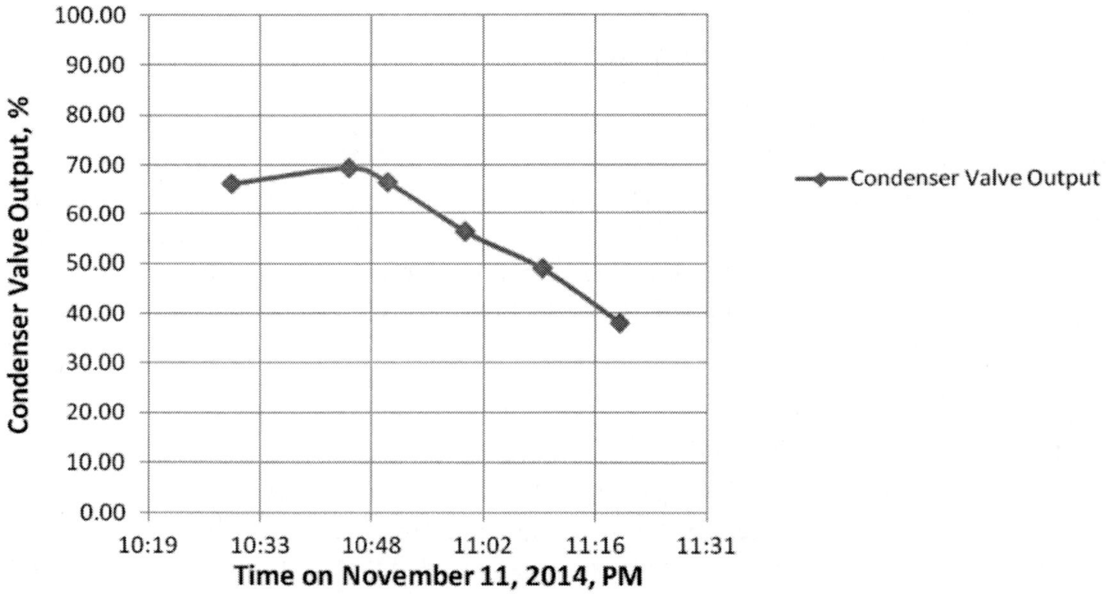

The operators noticed that since the restart, the system has operated much more smoothly. This too resulted from blanking the slots in the dip pipe. Prior to their blanking, the slots in the dip pipe poured subcooled liquid onto the liquid surface in the drum. This cold liquid reduced the vapor pressure in the drum. The amount of liquid pouring out of the slots was variable, changing with the pressure controller actions. The degree of subcool, therefore, would also vary. Varying amounts of subcooled liquid hitting the drum hot liquid surface would cause fluctuations to the drum pressure and therefore destabilize the system. Closing the slots stabilized the drum surface temperature and vapor pressure and therefore the drum pressure.

CONCLUDING REMARKS

Our experience teaches that when it comes to flooded condensers, "the devil is in the details." Adhering to good design practices, such as those described in reference 1, is a must. However, there are new things that we learn even in this well-established science, and experience is the best teacher. Event-timing analysis is a powerful tool for analyzing and acquiring the understanding that would pave the way for diagnosing flooded condenser control problems. With a correct diagnosis, minor equipment changes at negligible cost are often sufficient to provide total solutions and to reinstate trouble-free operation.

REFERENCES

1. Kister, H. Z. "Flooded Condenser Controls – Principles and Troubleshooting" , Chemical Engineering, January, p..37, 2016.
2. Kister, H. Z., "Distillation Troubleshooting", Wiley – Interscience, Hoboken, New Jersey, 2006.

481697 Tricky Liquid Distributors

Abstract:

Despite of continuing professional interest, liquid distributors selection and design are still challenging even for reputable vendors. Overlooking the basics often result in costly operational problems.

Case Study #1: An intermediate petrochemical product is manufactured using ammonia as the main feedstock. Unconverted ammonia is removed from the gaseous product by sulfuric acid absorption in the absorber column.

Absorber column has been revamped some years before with structured packing and new internals. After a period of reliable operation, absorber column performance suddenly droppedand unit throughput had to be reduced accordingly causing a substantial profit loss.

Field observation addressed liquid distributor as a problem cause, despite of dissent of tower internals vendor. A short shutdown was planned for column inspection and distributor repair. When the absorber column was openeda massive damage of packing was revealed. Surprisingly the column shell was also damaged as a consequence of pyrophorics self-ignition during the shutdown operations.

Investigation has shown that wrong distributor selection wasthe mainroot cause of the problem. New tower internals were provided on an emergency basis, including a new distributor of different type. Column shell was repaired in the vessel manufacturer's shop. Unit was successfully put into operation after a three weeks' repair period.

Case Study #2: Atmospheric distillation of crude oil is a core process of anyoil refinery. The technology is well known for many decades and equipment design had been standardized, apparently leaving no room for surprises.

After the startup of grass-root atmospheric distillation unita main fractionator column could not been operated at design conditions. When feed heater outlet temperature was increased to design condition,the columnwas shaking because of frequent hydraulic shocks in pumparound section. Hydraulic shocks could be prevented only at cost of valuable diesel fraction loss to bottom residue, keeping feed heater outlet temperature low.

Problem analysis addressed pumparound return distributor as a problem cause, which tower internals supplier could not believe. Column internals modifications were performed during a short shutdown. Unit was successfully started after the modifications and reached design capacity and product quality without any sign of hydraulic shocks.

481698 Two Phase Flow Feed Pipes to Distillation Columns - Some Quick Considerations

Abstract:

When the practicing engineer has to cope with two phase feeding lines, either in designing new installations or reengineering old ones, most of the time only incomplete or inaccurate information is available to justify the cost of performing a transient simulation analysis to assure a stable operation of the equipment been fed. The verification that the feeding line follows some simple and practical principles greatly minimizes the risk of ending with a pipe layout that generates additional flow instabilities to those already naturally occurring in the process.

Distillation Tower Inspection for Turnarounds and Revamps

185 Kister Distillation Symposium 2017

Doug Bouck
Technical Consultant
Separation Solutions, Ltd

Safety

- Personal
  - Basic Refinery/Chemical Plant Safety Training
  - Confined Space Safety Training
  - Hole Watch Communication
    - Voice or Radio (Compatibility Issue)
  - Know Your Own Limits
    - Heights
    - Closed In Spaces
    - Darkness
    - Flexibility
    - Conditioning

2

Safety

- Hazards
  - Climbing
    - Outside: Cluttered Platforms and Ladders
    - Inside: Cords, Hoses, Damage
  - Encumbrances & Hang-ups: Harness, Meters, Radio, Tools, Coveralls
  - Falling Objects: Tray Parts & Tools
  - Trays & Other Internals: Loose, Corroded, Displaced, Sharp Edges
  - Partner or Other Workers/Inspectors
  - Tight Spaces – Exit Plan

3

Safety

- Hazards
  - Dust, Scale
  - Old Equipment - Asbestos Gaskets
  - Pyrophoric Dust & Scale Producing SO2/H2SO3 Over 104F
  - Packing Fire Prevention:
    - "Causes and Prevention of Packing Fires", Chemical Engineering, p. 34 (July 2007).
- YOU are Responsible for YOUR Safety
- YOU are Responsible for OTHER'S Safety

4

5

Typical Tower Internal Manways

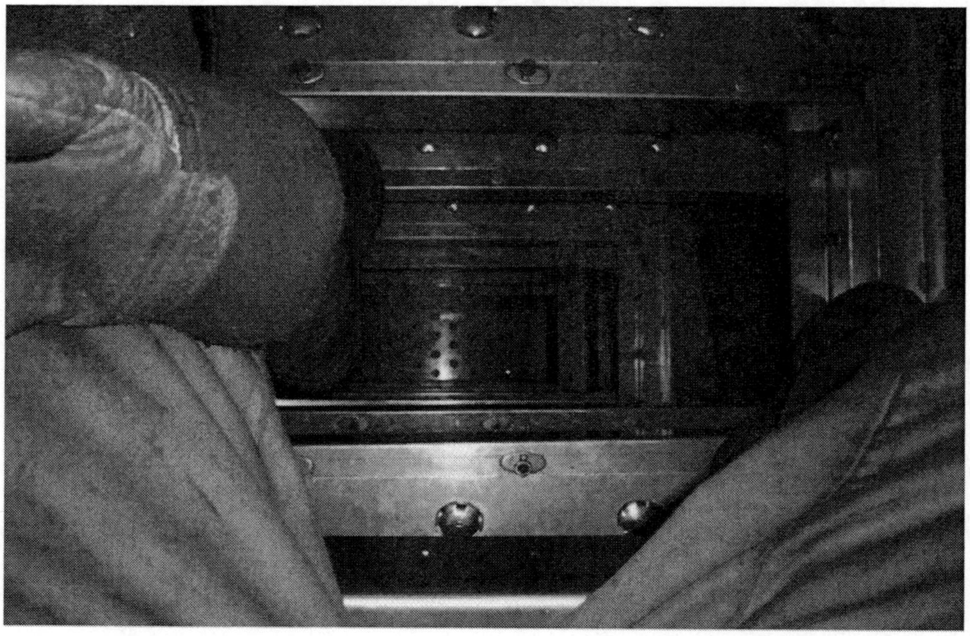

6

Tight Spaces

7

Multiple Personnel

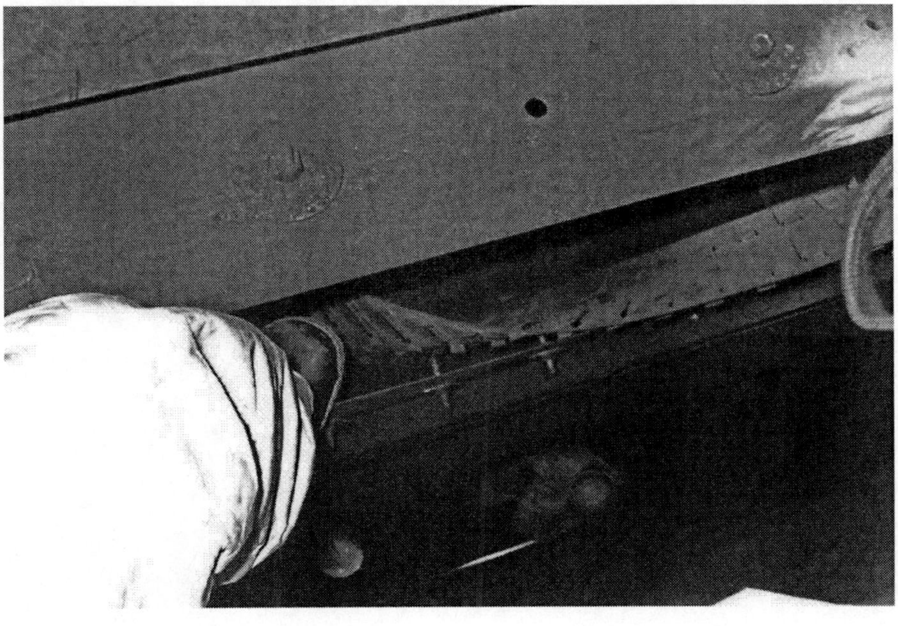

8

Tight with Tray Damage

9

Cluttered Manway & Internals

10

Scaffold Inside Tower

11

Workplan Responsibilities

- Process Inspection
- Maintenance/Mechanical Inspection
- Vendor Reps./Inspectors
- Contractor Inspectors
- Operations Inspection
- Training Inspection Tours
- Punch Lists and Change Orders
- Follow-up Inspections
- Signoff Responsibilities

12

Inspection Equipment

- Typical in Waist Pack
  - Lock-out Lock and Tag
  - Zip Lock Bags & Magnet
  - Soapstone + Yellow Crayon + Black Marker
  - 10-25 ft Tape Measure
  - 9/16 Wrench & Plyers
- Other
  - Shin Guards and/or Knee Pads
  - Note Pad, Drawings & Inspection Plan
  - 2 Light Sources with Fresh Batteries
  - 2+ Pens/Pencils
  - Camera (Hot Work Permit)

13

Waist Pack Contents

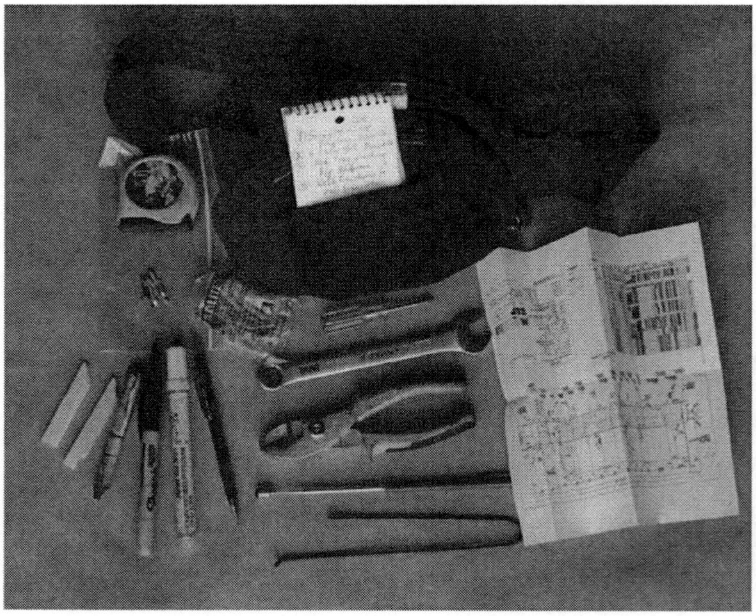

14

Other Equipment

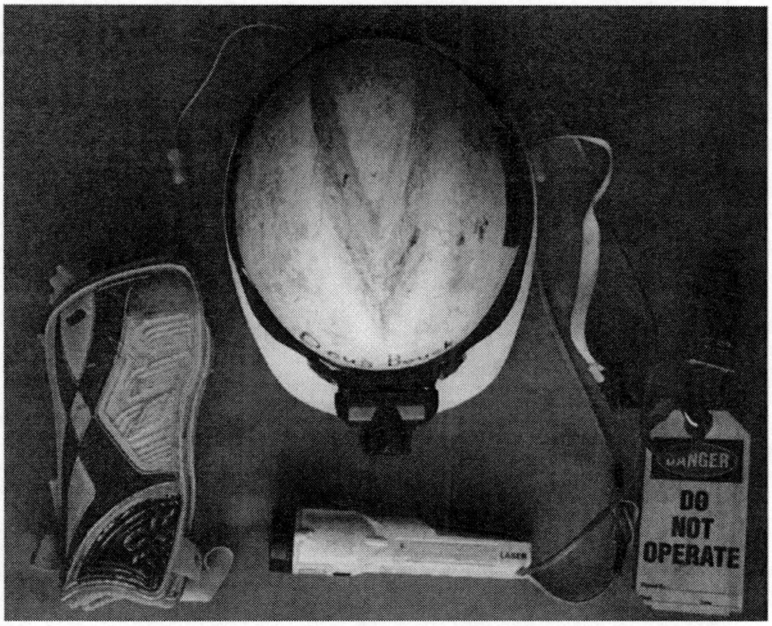

15

Inspection Equipment

- Special
  - Goggles, Dust Mask or Face Mask
  - Weir/Downcomer Dimension Template
  - 6 Inch Steel Rule
  - Dimension & Tolerance Table
  - Detailed Tray/Internals Drawings
  - Punch List and/or Drawings
  - Level Device: Bubble, Water or Laser
    - Device Accuracy Test
  - Distributor Test Equipment
  - Water Tightness Test Equipment

16

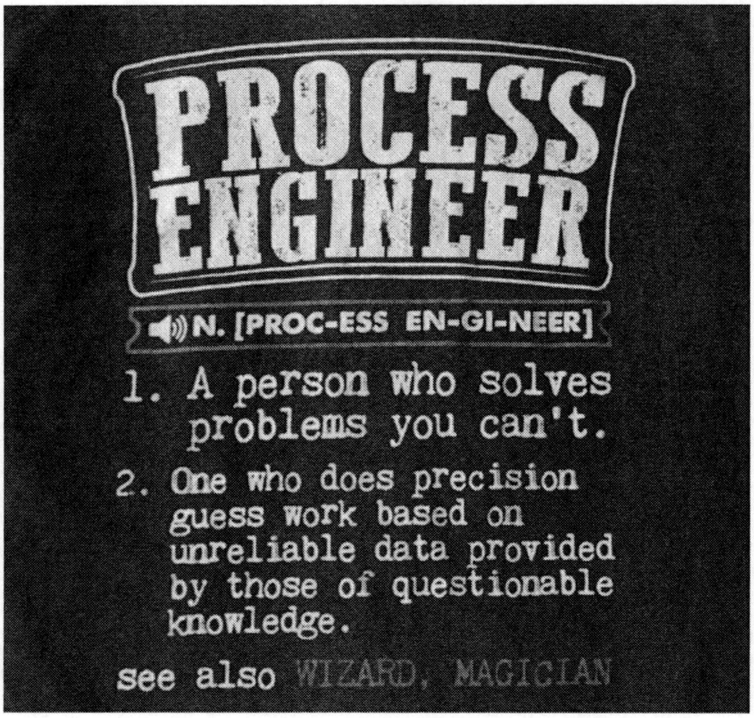

17

Keys to Process Inspection

- Understand How the Process Works
- Understand How the Parts Fit Together
- Understand How the Parts Work Together
- Know Why the Specs are Important
- Know How Deviations Affect the Process
- Be Able to Explain Operating and Economic Effects of Deviations
- Consider the Next Fix Chance May be 5 Years
  - Know When to Hold'em and When to Fold'em

18

Basic Preliminary Tray Inspection

- Number Trays for Reference
- Note Cleaning Issues – Take Photos
  - Does It Need Cleaning Before Full Prelim.?
  - Loose Deposits or Hard Scale?
  - Tray Decks/Downcomers and On/Under Valves
  - Any Special Cleaning Techniques Required?
- Note & Mark Missing/Stuck Valves
- Note & Mark Loose or Missing Hardware
- Note Cracked/Damaged/Corroded Trays
- Note Condition of Internal Manway Hardware
- Note Condition of Feed Pipes & Vortex Breakers
- Prepare Recommendations & Punch List

19

TAR Tray Inspection 1

- Objectives:
  - Return Equip. to Efficient Operation
  - Remember the Next Fix Opportunity May Be Years

- Cleanliness
  - Loose Scale/Deposits on Downcomers & Walls
    - Any Loose Scale Can Move to Downstream Equipment
  - Deposits on Decks
    - Plugging or Restriction of Valves, Holes or Caps
    - Bubble/Tunnel Caps Annular Space Must Be Clear
  - Accumulation in Downcomers, Sealpans & Drawpans

20

TAR Tray Inspection 2

- Missing/Worn/Stuck/Plugged Valves
  - One (1) Percent Acceptable Exception
  - Moving Valves Must Move Full Travel
    - Remove Deposits on Top, Legs or Around Holes of Valves
  - Check for Worn Valves & Tray Decks
  - Replacement Valves: Legs Bent for Retention
  - Replacement Valves with Cages: Crimping of Leg Tabs

21

Valve Cage Installation Problem

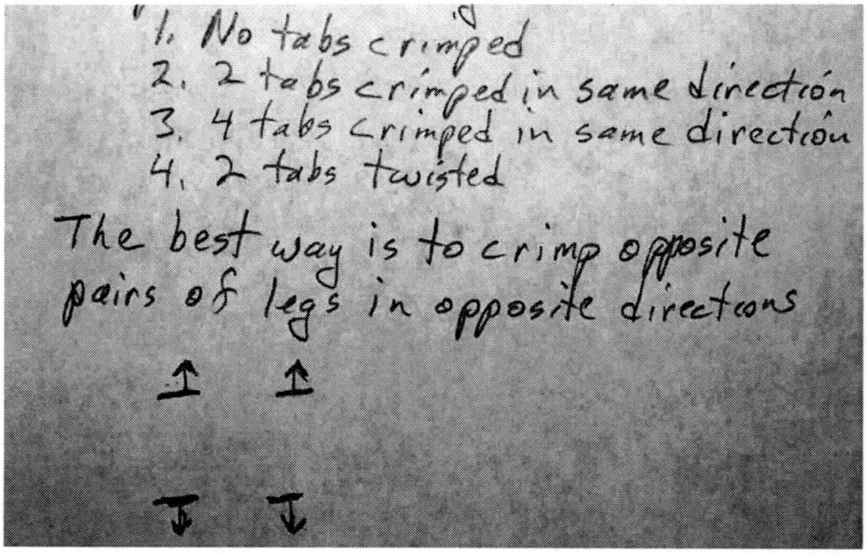

22

TAR Tray Inspection 3

- **Most Frequent Problem:**
  - Frozen/Damaged Internal Manway Hardware
    - Result of Hammer Release and/or Corrosion
    - Frozen Hardware Guarantees Poor Re-Closure
    - Poor Closure Causes Vibration Cracking of Trays
    - Poor Closure Causes Less Security Against Upset
    - Poor Closure May Result In Operating Deficiency
  - **Hardware Problems Must be Corrected Before Approval for Internal Closure**

23

TAR Tray Inspection 4

- Nutter MW Hardware Problems:
  - Subtle Correct Orientation
  - Rectangular Slots and Bolts
  - Boilermaker Hammer Release Damage
  - Same Poor Closure Issues
- Blind Fastening
  - Unusual
  - Difficult
  - Two Different Types of Hardware
  - Two Different Techniques
  - Boilermakers May Need Instruction

24

Nutter Internal Manway Hardware Orientation

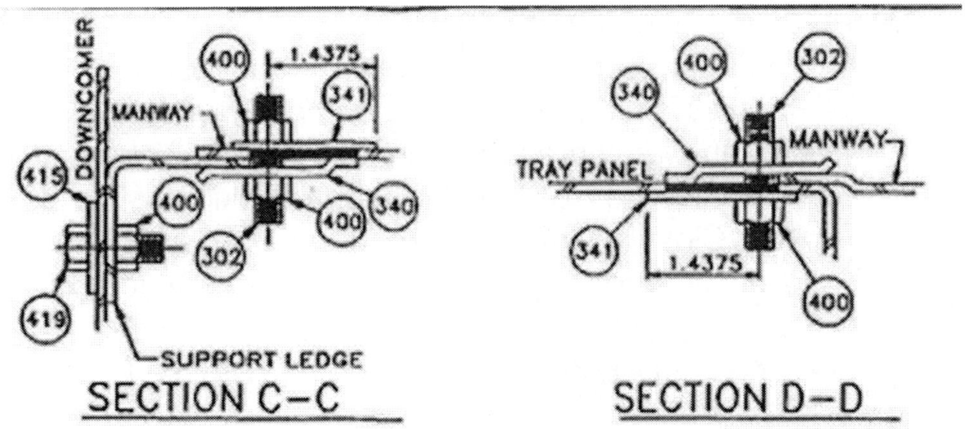

25

Blind Manway Bolting

26

Blind Manway Bolting

27

TAR Tray Inspection 5

– Other MW Hardware Problems
  - Light Wt. Football Washer Bends Causing Poor 2nd Closure
  - 3/16 in. Thick Heavy Duty Washers (Footballs) Recommended

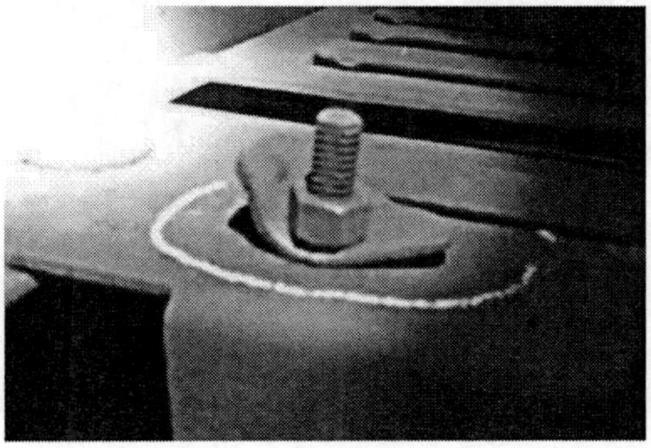

28

Holes in Downcomers

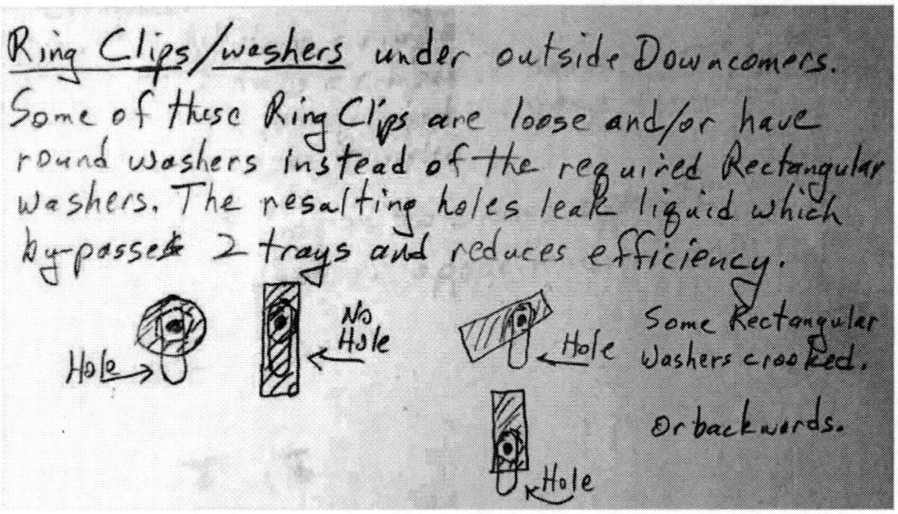

29

Internal Downcomer Loose Scale

30

Downcomer Scale

31

Dirty Round Valve Tray

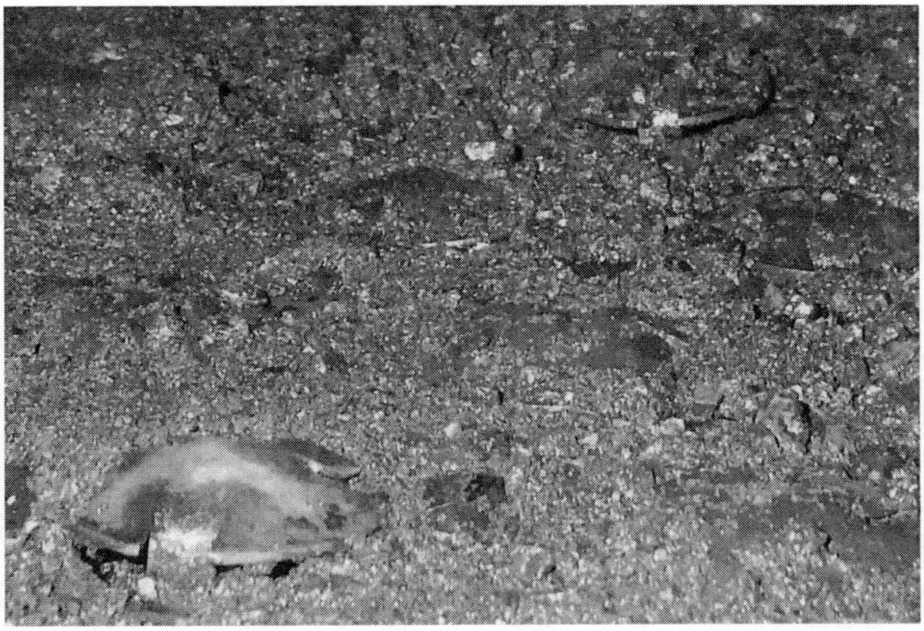

32

1- 1.5 Inch Deposits

33

Corroded and Plugged Bubble Caps

34

Displaced Weir

35

Bent/Loose Downcomer Clip

36

Bent Tray Downcomer & Clip

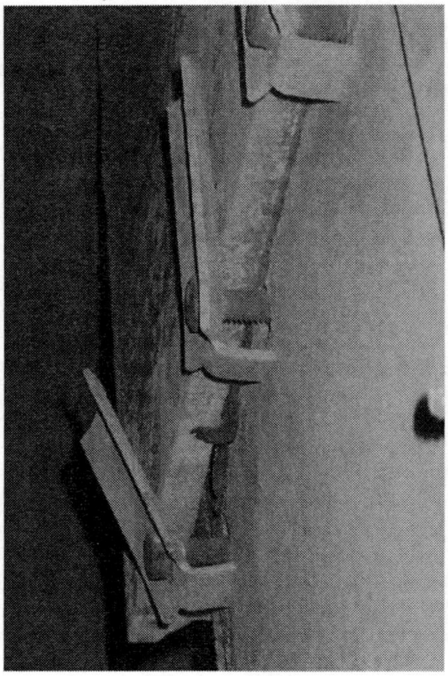

37

Ring Clip Askew

38

Displaced Valves

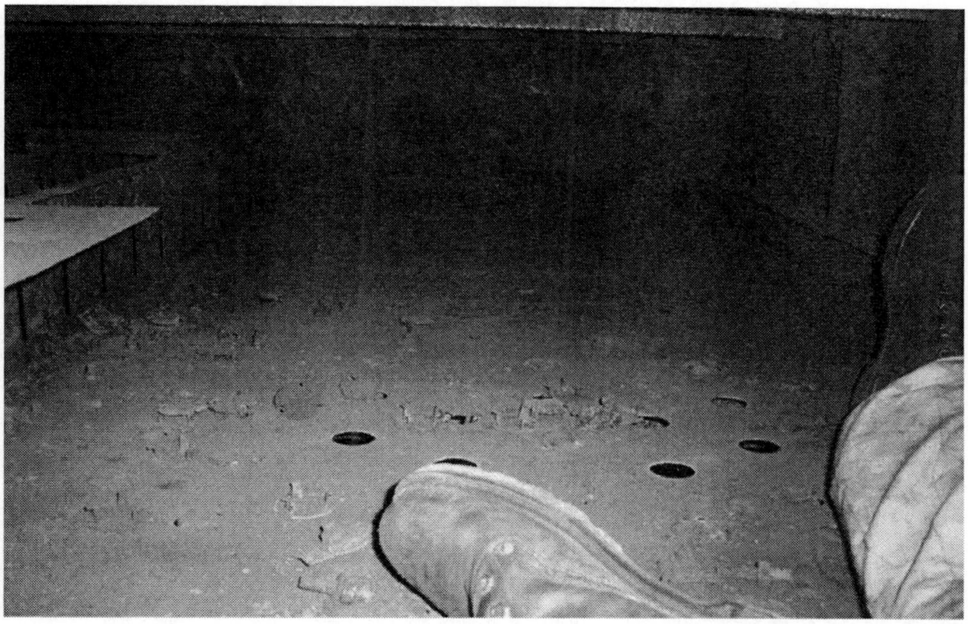

39

Displaced Valves

- Check the Legs Condition
  - Valve Legs Worn?
  - Improper Installation?
    - Legs not Spread Properly?
- Check the Hole Condition
  - Worn By Erosion-Corrosion?
- Blown Loose by Process Problem?
- Deck Thinning from Corrosion?

40

Missing Bubble Cap

41

Missing Swept Back Weir Piece

42

Missing Deck Bolts

43

Tray Damage Analysis

- Trays Bent or Cracked
  - Fatigued by Age and/or Weight
  - Overtightening
  - Level Loss or Explosion
- Trays Pushed Up
  - High Level + Percolating Vapor
  - Water Explosion Below
- Trays Pushed Down
  - High Level + Rapid Level Loss
  - Water Explosion Above

44

Boss says:

DOUGBERT, THE CEO SAYS THAT THE PROBLEM DIB COLUMN IS OPEN FOR INSPECTION AND YOU ARE THE MAN!

Dougbert says:

WHY ME? THAT TOWER IS OVER 200 FEET TALL!

Boss says:

HE SAYS YOU ARE THE ONLY ONE WHO KNOWS HOW THE INSIDES ARE SUPPOSED TO WORK.

45

Bent Tray Panel at Support Ring

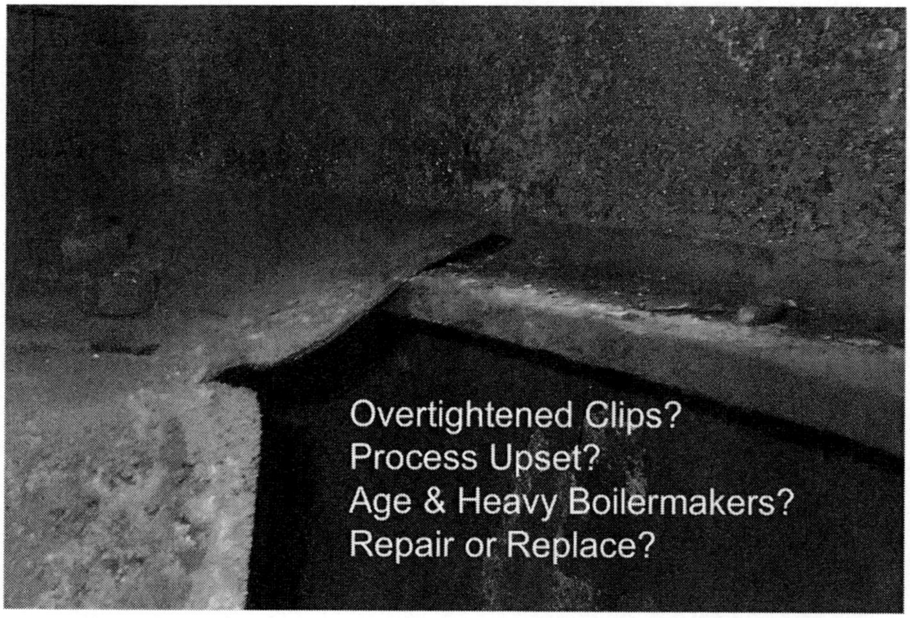

46

Results of Loss of Level Control with Vapor Pushing Liquid Up Through the Trays

47

Uplifted Downcomer Panel

48

Lift Damage

49

Lift Damage

50

Lift Damage

51

Push Down Damage

52

Push Down Damage

53

Brittle Failure

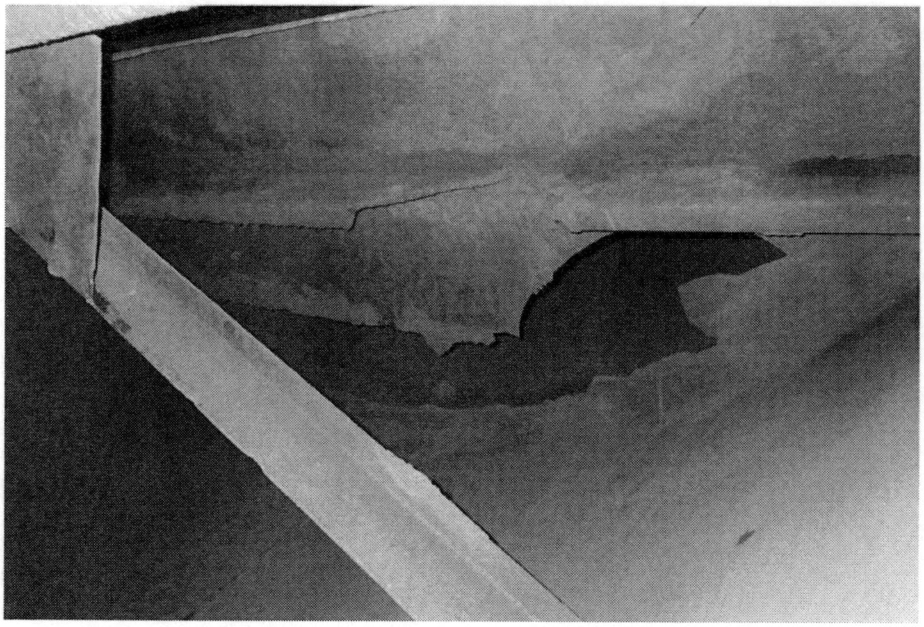

54

Process Revamp Review

- Pre-TAR Revamp Process Review
  - What is Different
  - Objective of Change
  - Impact of Change on the Process
  - Will the Changes Meet the Objectives
  - Are the Changes Consistent and Valid Engineering?
- Pre-TAR Mechanical Review
  - Mechanical Design meet Process Design?
  - Integration with Surrounding Systems: High Capacity Trays + Reboiler + Condenser?

55

Revamp Inspection

- Installed Mechanical & Process Check:
  - Installed per Drawings
  - DC Installation Blocks (Record DC Escapes & Any DC Changes)
  - Alignment of Tray Sections (Centering & Panel to Panel) Out of Round?
  - Correct Hardware Installation
  - Hardware Orientation & Proper Tightening
  - Punch List & Final Process Check

56

Revamp Field Inspection Examples

- Parts Verification Check: Example Valve Lift
- Small Parts Check and Security: Examples Twr. Closure Hold-up & Missing Nozzles
- Field Process Check: Example Missing Downcomers
- Field Dimension Check: Example Stripper Dbl. Flood
- Field Dimension Check: Support Ring Thickness
- Inspection Timing: Example Redesigned Feed Pipe
- Field Layout: Example Chimney Tray Water Draw

57

Example: Chimney Tray Water Draw

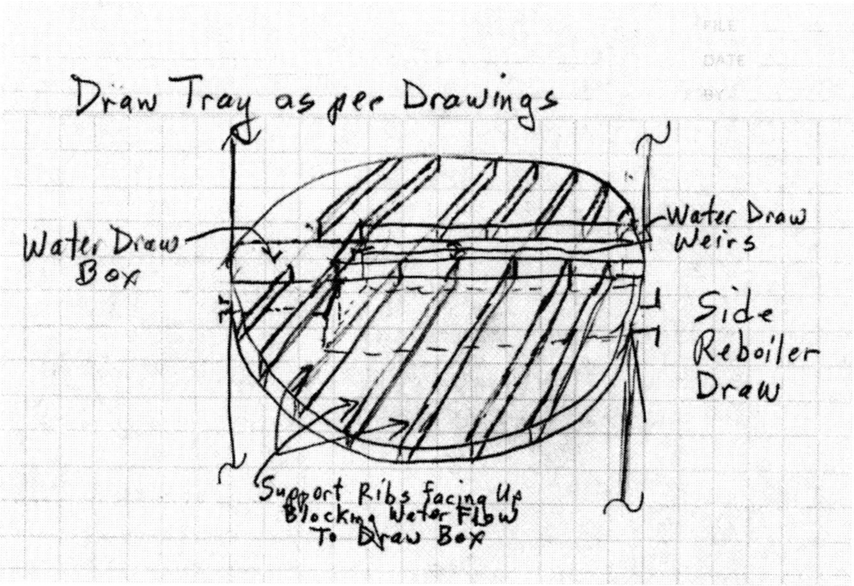

Bouck, Doug, Distillation Revamp Pitfalls to Avoid, Chemical Engineering Progress, p. 32-38, Feb. 2014

58

Special Inspections 1

- Feed and Return Pipe Distributors
  - Holes/Slots/Nozzles
    - Clear?
    - Corroded?
  - Flange Bolts Tight (Wrench Check)
  - Distributor Support Bolting Tight (Dbl. Nutting Use)
- False Downcomers & Distributor Impact Panels
  - Double Check for Loose Hardware
  - Need for extra security from vibration?
  - Vibration initiated cracking?

59

Distributor Hole Corrosion

60

Special Inspections 2

- Material Balance or Draw Problems?
- Flow or Level Control Problems?
- Suspected Leaks?
- Chimney Trays & Draw Boxes
  - Corrosion & Deposits in Draw and/or Piping
  - Thickness Checks in Suspect Corrosion Areas
  - Potential Leak Areas at Welds or Gasketing
  - Potential Leak Areas at Bottom (Under-deposit Corrosion)
  - Loose or Bent Hats, Chimneys or Decks

61

Chimney Draw Tray

62

Draw Box Corroded Through at Bottom

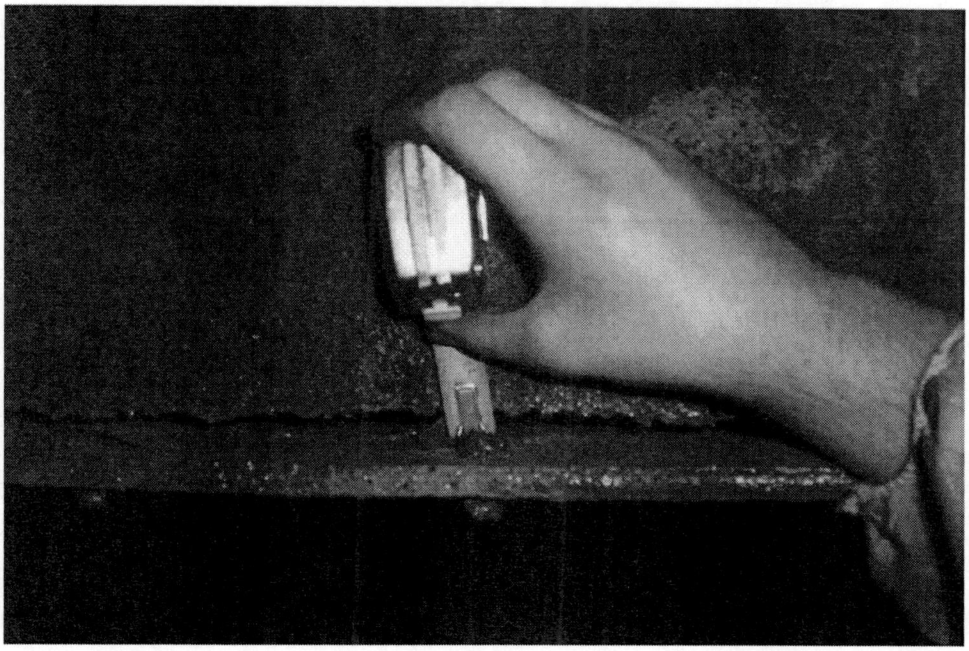

63

Draw Box Corroded Through at Welds

64

Special Inspections 3

- New or Revamped Chimney Trays
  - Gasketed or Welded
  - Leak Testing
    - Water Quality (Chlorides for Certain 300 Series Stainless and/or Solids in Firewater)
    - Quantity - Multiples Greater Than Leak Rate
    - Pressure - Pump Head Requirements (Water)
    - Test Depth - Approx. Operating Level
- Distributor Testing
  - On Site Spray Distributor Example
  - Off Site Packing Distributor Vendor Testing

65

Special Inspections 4

- Final Inspection Before Closure
  - All Internal Repairs & Cleaning Complete
  - Internal Manway Hardware Fixed or Replaced
  - Account For All Internal Manways – Directional?
  - Tools & Misc. Picked Up

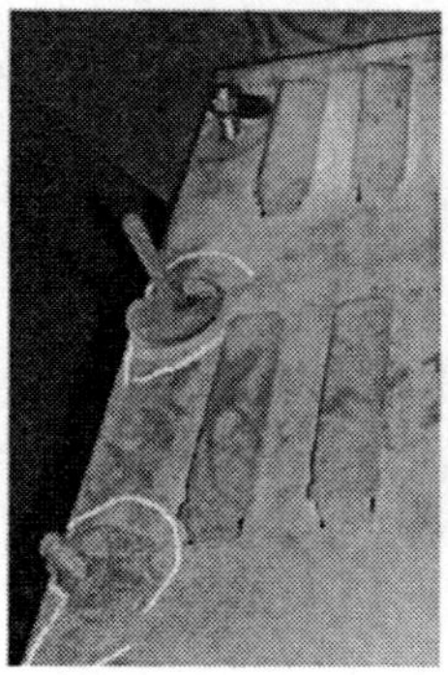

66

New Tower Inspection 1

- Tolerances
- Level Checks
  - Trays – Dependent on ring levelness
  - Weirs (If Adjustable)
- Downcomer Width – Top & Bottom
- Downcomer Escape
- Weir Heights (If Adjustable)
- Alignment/Centering of Tray Sections
- Valve Rise & Valve/Hole Counts

67

New Tower Inspection 2

- Hardware
  - Per Drawing
  - Correct Metallurgy
  - Bolting Snug Tight (~15 ft-lbs for 3/8 inch)
  - NOT Overtight
    - Bend Tray
    - Bend Clamps
    - Bend Washers
    - Gall Bolts

68

Pyrophoric Fire Prevention

- Pyrophoric Deposits
  - Iron Sulfide Can Be Most Anywhere Sulfur Is
  - Conversion Treatments Not Perfect
  - Prefer One-Step Processes Like ZymeFlow
  - Preventive Measures + Monitoring
  - FeS + O2, Dry and 104+ Degrees F. = Exothermic
  - Examples: Amine, Main Fractionators
- Carbon, Coke & Other Organics
- Hot Work
  - Read Chemical Engineering July 2007 Article on "Causes and Prevention of Packing Fires"

69

Packed System Inspection with NO Packing Removal

- No Specific Operating Problems
- Pre-TAR Gamma Scan
- Distributor Piping – Clean & Bolting Tight
- Pre-Distributor - Clean & Functional
- Distributor Pan, Troughs or Spray
  - Clean
  - Holes or Nozzles Clear
- Visual Inspection Top of Packing
- Visual Inspection Bottom of Packing

70

Structured Packing Installation

Feed Inlet Pipe
Predistrubutor
Distributor
Optional Holddown
Top Layer of Packing
90 Deg. Oriented Next Layer

Alternating Orientation of Layers

Bottom Layer of Packing
Packing Support
Liquid Collector

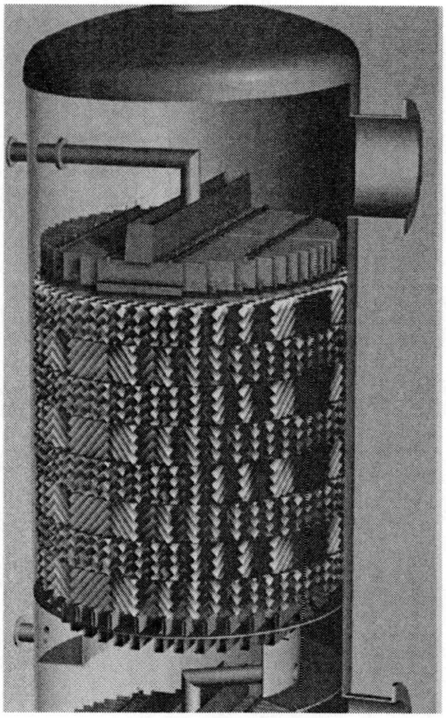

Sulzer Chemtech Brochure

71

Packed System Inspection With Packing Removal

- Before Re-Installation of Packing Check:
  - Draw Integrity – Check for Corrosion & Cleanliness
  - Support Pan/Grid – Clean/Unplugged
  - Check Support Orientation
  - Check Connections and Support Bolting Tightness
- Structured Packing Installation:
  - Verify Packing Size, Type & Metallurgy
  - Starting Orientation is Important
  - 90 Degree Alternate Layers (45 or 60 Degree Rotation for Grid)

72

Protect the Packing

- Walk on Planks NOT the Packing
- WHY for Structured Packing??
  - Structured Packing is Very Thin
  - Boots Smash the Top Crimps
  - Smashed Crimps Reduce Capacity
  - Smashed Crimps Reduce Efficiency
  - Hidden Damage is Untraceable
- Make Sure the Contractor is Complying

73

Protect the Packing

Stand on Planks
not the Packing

74

Protect the Hands/Arms

Koch-Glitsch Brochure

75

Installation

- Personal Safety
  - Gloves for Sharp Edges
- Orientation to Support
- Orientation Layer to Layer
- Shoehorns for Multi-piece Installation
- Distribute Excess Space Between Pieces
- Wall Wipers of Various Designs

76

Structured Packing Layers

FLEXIPAC® Structured Packing with perforated and textured surface shown with attached wall wiper band

77

Other Wall Wipers

Sulzer Mellapac Structured Packing

78

Holddown Grid – Miss-installed

79

Distributor Inspections

- Holddown – Bolting Tight
- Level Distributors & Pre-Dist. 1/8 in. High to Low
- Installation Hold Point for Level Check/ Adjustments
- All Distributor and Predistributor Bolting Tight
- Distributor Clean
- Predistributor Clean
- Feed Piping Re-installed & Bolted Tight

80

Spray Distributors

- Check all Spray Nozzles and Headers
  - Remove Nozzles (Have Spares on Hand)
  - View Nozzles Internals – Clean or Replace
  - Save Nozzles in Safe Place
  - Check and Clean Headers as Necessary
  - Wrench Check all Nozzles for Tightness
  - Wrench Check all Flange Bolting
  - Double Nut and Check all Header Support Bolting

81

Detached Spray Headers

82

Upset Structured Packing (No Holddown Grid)

83

Redistributor Between Packed Beds

84

Random Packing Installation

Feed or Transfer Pipe
Predistributor
Distributor
Retaining Grid

Random Packing

Support Plate
Support Ring

Vapor Inlet

Liquid Level Collector/Sump

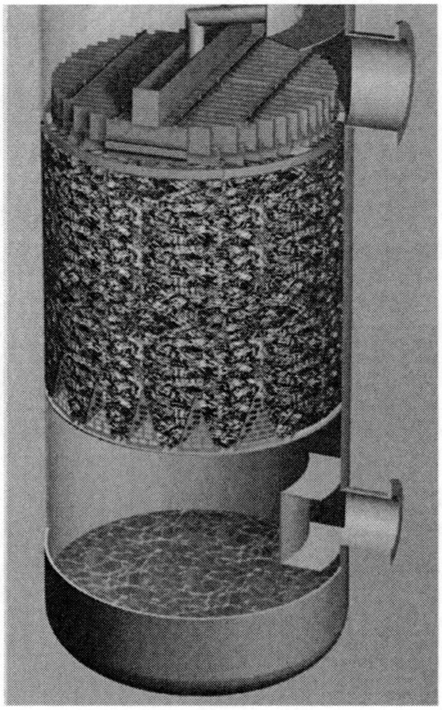

Sulzer Chemtech Brochure

85

Packing - Random

- Removal for Replacement
  - Remove Piping, Distributors & Retaining Grid
    - Safe Storage of All Parts
    - Record Orientations for Re-Installation
    - Clean All Parts
  - Remove Rings by Vacuum
  - Clean Any Collector or Draw System
  - Clean Packing Support
  - Check Support Integrity & Bolting

86

Packing - Random

- Installation
  - Verify Packing Size, Type and Metallurgy
  - Dump New Packing
  - Rake Top
  - **Do Not Overfill!**
  - If Necessary to Enter, Stand on Planks not Packing
  - Re-install Retainer Grid
  - Level Distributors & Pre-Dist. 1/8 in. High to Low
    - Installation Hold Point for Level Check/Adjustments
  - Install Feed Piping
  - Check Tightness of All Bolting

87

Protect the Packing

- Walk on Planks NOT the Packing
- WHY for Random Packing??
  - Random Packing is Thin
  - Boots Crush or Compress the Rings
  - Hidden Damage is Untraceable
  - Crushed Rings Reduce Capacity
  - Crushed Rings Reduce Efficiency
- Make Sure the Contractor is Complying
- Stay Totally Off, If Possible

88

The Process Engineer
(A Bit of Philosophy)

- Little Authority
- Much Responsibility
- Bridge Operations, Maintenance & Management
- Optimization Economics - Beyond Politics
- Relationships Create Influence
- Influence Creates Accepted Recommendations
- Recommendations Create Improvements
- Improvements Create More Profits
- More Profits Create Stable Company and Jobs

89

Take Ownership of the Future. The Role of Process Engineers in Inspection **Is** Important!

2 Peter 1:15

And I will make every effort to see that after my departure you will always be able to remember these things.

YOU ARE the FUTURE!

90

481654 Distillation Tower Inspection for Turnarounds and Revamps: Tutorial and Presentation - Part II

Abstract:

This presentation of "Distillation Tower Inspection for Turnarounds and Revamps" covers the unique safety issues for Distillation Tower inspections plus typical work plan responsibilities and inspection equipment. Both preliminary and follow-up inspection of trays and packed systems are covered. Emphasis is placed on what to look for in typical and unique inspection situations with both trayed and packed towers. Examples will be demonstrated and/or discussed including the unique perspective that the Process Engineer brings to the inspection process especially in replacement and revamp situations.

478993 Quantification of the Effect of Viscosity on Point Efficiency of Sieve Trays in Distillation Columns

Abstract:

Though many physical properties govern the design and performance of distillation columns, the effect of viscosity is critical as it influences both mass transfer and hydrodynamic aspects of distillation columns. While there has been research to analyze the effect of viscosity on distillation, the majority of the research was conducted with liquid viscosities less than 1.0 cP. The distillation efficiency correlations based on low viscosity data may not be reliable for systems that exhibit liquid viscosities more than 1.0 cP at operating conditions. Point efficiency data collected using Oldershaw columns are proven to be effective in representing the commercial point efficiency data. Therefore, an Oldershaw column was recently commissioned at OSU to collect efficiency data for liquids with elevated viscosities. Column validation is done using standard test systems for which commercial efficiency data are readily available. The efficiency data at elevated viscosities (up to 4.0 cP) are collected by Oldershaw efficiency experiments at total reflux. The collected data are crucial in order to evaluate the validity of point efficiency correlations at high viscosities. Also, the elevated viscosity data are critical for extending existing correlations and developing new correlations for viscous systems.

479217 Investigating the Impact of Feed-Air Saturation in Air-Water Spray Entrainment

Abstract:

Entrainment characterization is critical for a variety of separation processes in the chemical process industry, including distillation and absorption. Liquid entrainment is a key aspect of these applications that is not well understood. In order to support further research into this field, it is important to determine the correct methods by which to analyze entrainment in laboratory systems. This work investigates the impact of saturating the feed air to an air-water entrainment system, particularly on the droplet size and velocity distributions as measured by Phase Doppler Interferometry. Entrainment measurements were made at the same experimental conditions with and without pre-saturating the feed, and the results were compared to determine the impact on those measurements. These method evaluations are essential as they guide the larger research effort in correlating entrainment with efficiency and capacity.

Effect of Liquid Viscosity on Mass Transfer for Packings

Di Song[a,b], A. Frank Seibert[b], Gary T. Rochelle[a,]*

[a]*McKetta Department of Chemical Engineering, The University of Texas at Austin, 200 E Dean Keeton St., C0400, Austin, TX 78712, USA*

[b]*Separations Research Program, Pickle Research Center, The University of Texas at Austin, 10100 Burnet Rd., Austin, TX 78758, USA*

Abstract

It is important to understand the effect of liquid viscosity (μ_L) on both the effective mass transfer area (a_e) and liquid film mass transfer coefficient (k_L) for packings to correctly predict the separation rate and properly design the column. Existing correlations are mostly based on water system and the application of them to viscous media could be problematic.

In this work, a_e and k_L were measured in a semi-works scale packed column (16.8 in. I.D.) with 10 ft packings installed as the internal. Seven structured packings (M 125Y, M 250Y, M 250X, GTO 250Y, B1 250MN, GTP 350Y, GTP 500Y), one random packing (RSR 1.5), and one hybrid (RSP 250Y) were tested. Liquid viscosity was varied from 0.8 to 70 cP by adding glycerol to water. The a_e was determined by reactive absorption of ambient CO_2 into dilute NaOH, and the k_L was determined by air stripping of toluene from the liquid. Liquid viscosity showed insignificant effect on a_e over the range of μ_L investigated. The total dependence of k_L on μ_L is -0.72, of which -0.35 is from the indirect influence through diffusivity, and -0.37 is from the direct influence of μ_L on k_L through liquid turbulence.

Models of a_e (Eq. 1) and k_L (Eq. 2) were developed based on the nine measured packings together with other 18 packings in the database of the Separations Research Program (SRP). The effect of secondary mass transfer (wall and end effect) has been included in the models.

$$\frac{a_{e,packing}}{a_P} = 1.26 \cdot \eta \cdot \left(We \cdot Fr^{-\frac{1}{3}}\right)^{0.117} = 1.26 \cdot \eta \cdot \left[\left(\frac{\rho_L}{\sigma}\right) g^{1/3} \left(\frac{u_L}{a_P}\right)^{4/3}\right]^{0.117} \quad (1.a)$$

$$a_{e,total} = a_{e,packing} + a_{secondary} \approx a_{e,packing} + a_{wall} \quad (1.b)$$

$$Sh = 0.124 \cdot Sc^{0.5} \cdot \left(Re \cdot Ca^{1/5} \cdot Fr^{-1/3}\right)^{1.08} \quad (2.a)$$

$$k_L = 0.124 \cdot u_L^{0.576} \cdot \mu_L^{-0.364} \cdot D^{0.5} \cdot a_P^{-0.44} \cdot g^{0.36} \cdot \rho^{0.58} \cdot \sigma^{-0.216} \quad (2.b)$$

With a relatively large equipment size and packing database, this work has the greatly expanded the data of mass transfer in viscous liquids available in the open literature.

Introduction

Many industrial applications encounter mass transfer in viscous liquid, such as electrolyte solutions, organic/polymer solutions, crude oil, and ionic liquids, etc. The increase in μ_L can significantly decrease k_L. Take post-combustion CO_2 capture with aqueous amine for example, the high solvent viscosity can cause slower diffusion of CO_2 to the bulk liquid, slower diffusion of free amine from the bulk liquid to the gas-liquid interface (surface depletion of alkalinity), slower diffusion of loaded amine back to the bulk liquid (which affects equilibrium partial pressure of CO_2 at the interface), and less liquid turbulence on the packing surface. Most of the existing k_L (k_La) correlations investigate only water and thus provide unsatisfactory predictions for the influence of μ_L [1]. Therefore, the application of them to amine solutions could lead to either inefficient or insufficient column design.

The SRP database for the air-water column contains valuable hydraulic and mass transfer data for various structured, random, and hybrid packings. Based on the database, models of a_e and k_L have been developed previously by Tsai [2] and Wang [3], respectively. The Tsai area model was developed seven years ago, and new data with totally different packing geometry have been generated continuously. For the k_L model developed by Wang, though it is relatively new, a large amount of new data with water and aqueous glycerol was generated during the past two years. Therefore, it is necessary to update both models with the expanded database.

Methods and Equipment

A systematic method for packing characterization with water has been developed at the Separation Research Program in the University of Texas at Austin. The a_e has been measured with chemical absorption of ambient CO_2 into aqueous NaOH because of the well-established kinetics [4]. The k_L has been measured with air stripping of toluene from water.

Glycerol was chosen as the liquid viscosity enhancer for its complete and easy dissolution in water, relatively cheap price, and the Newtonian behavior of its aqueous solution. Liquid viscosities of 65, 15, 4, and 1 cP (approx.) were obtained by adding 85, 65, 44, and 0 wt % glycerol to water. The reaction kinetics for CO_2/NaOH/H_2O/glycerol was calculated from a model developed based on the wetted-wall column data [5]. The glycerol will not change the controlling mechanism for mass transfer in the existing water systems. For toluene air stripping, though the infinite dilution activity coefficient of toluene decreases rapidly with increasing glycerol, the value will still be well above 100 at the highest proposed glycerol, which ensures the liquid film domination. For the reactive system, the effect of physical liquid film mass transfer (i.e. surface depletion of alkalinity) has been corrected in the kinetic model and is believed to be not an issue for the packed column experiments considering the relatively low CO_2 concentration and flux.

Packed column

The PVC packed column used in this work is sketched in Fig. 1. The inside diameter of the column was 16.8 in. The maximum packing height was 10 ft. The gas and liquid had countercurrent contact inside the column with air supplied by a 30-kW blower with variable speed drive from a duct (8 in outside diameter) below the packing support. The liquid was discharged by a centrifugal pump regulated with a variable speed drive with a maximum capacity of 150 gpm from a 350-gallon storage tank. The liquid could either flow to the column top and distributed by an F40 distributor (40 drip points/ft^2) or flow back to the column sump and storage tank via a bag filter for the pump to operate at a more favorable drive-speed region. Above the distributor was a Trutna collector to knock out any liquid reaching the column exhaust. Typical liquid inventory size was 250-300 gallon. Typical gas and liquid rates were 150-750 acfm and 1-30 gpm/ft^2 respectively.

Approximate 10 ft packing was installed for a_e measurement and k_L measurement with aqueous glycerol. A shorter 6 ft bed was used for k_L measurement with water because of the difficulty of GC analysis for the low outlet toluene concentration caused by high mass transfer efficiency. Each element was installed with 90° rotation with adjacent elements to facilitate liquid distribution as a routine practice.

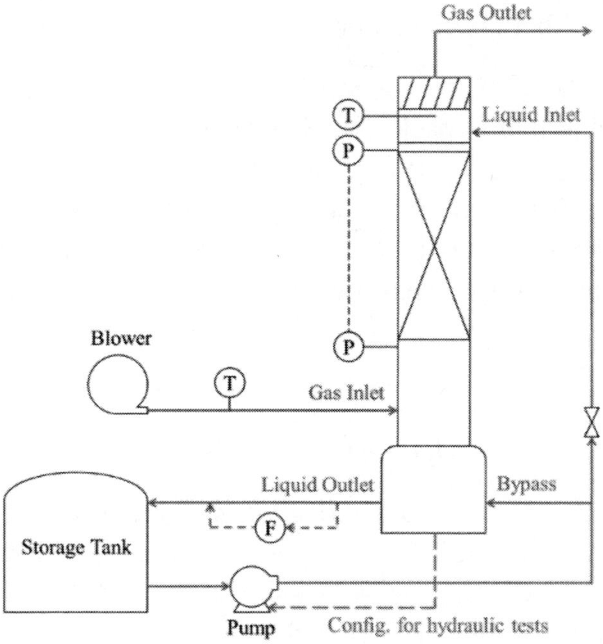

Fig. 1: Packed column experimental system.

Auxiliary Equipment

The CO_2 concentration in both the inlet and outlet gas was measured online by a XStream® CO_2 analyzer (Rosemount). Inlet CO_2 was ambient 370-400 ppmv, and outlet CO_2 was 100-300 ppmv depending on the packing and gas flow rate. Toluene concentrations of the liquid samples were determined by gas chromatograph (Hewlett Packard 6890) analysis with 1-bromo-4-fluorobenzene used as the internal standard. Typical inlet toluene was 100-200 ppm$_w$, and outlet being 0.5-50 ppm$_w$ depending on the liquid flow rate and viscosity. A viscometer (TA Instrument AR-2000ex) was used for liquid viscosity measurement. A densitometer (Mettler Toledo DE40) was used for density measurement to determine glycerol concentration for reaction kinetic calculations.

Calculations

The total mass transfer area in the column, $a_{e,total}$, was determined experimentally by Equation 3. The k_g' for aqueous NaOH was calculated using the model of Pohorecki and Moniuk [4]. The k_g' for caustic glycerol solutions was calculated from a model developed based on the wetted-wall column data [5]. The k_La_e was determined experimentally by Equation 4.

$$a_{e,total} = \frac{K_{OG}a_e}{K_{OG}} \approx \frac{K_{OG}a_e}{k'_g} = \frac{u_L}{k'_g Z}\ln\left(\frac{C_{CO_2,\ in}}{C_{CO_2,\ out}}\right) \quad (3)$$

$$k_L a_{e,total} = \frac{u_L}{Z}\ln\left(\frac{C_{toluene,\ in}}{C_{toluene,\ out}}\right) \quad (4)$$

Physical properties (density, viscosity, surface tension, diffusivity, Henry's constant of CO_2) of the aqueous glycerol solutions were calculated using models developed previously [1].

The secondary effect (end and wall effect) on mass transfer area ($a_{secondary}$) has been corrected to give effective mass transfer area of packing surface ($a_{e,\text{packing}}$). To calculate the wall area, it has been assumed that the column wall was totally wetted not only in the packing section, but also in the auxiliary section (for this case with only one packing section, it's only the section between packing bottom and liquid sump):

$$a_{wall} = \pi \cdot D_{column} \cdot (Z_{packing} + Z_{auxiliary}) \quad (5)$$

The top area includes area generated by the liquid streams from distributor to packing top (both stream peripheral and impact area). Similarly, the bottom area includes area generated by the liquid streams from packing bottom to sump. It has been demonstrated in later section that wall area is always dominant in $a_{secondary}$.

Model of Mass Transfer Area (a_e)

The $a_{secondary}$ has been deducted from $a_{e,total}$ to give $a_{e,\text{packing}}$ (Eq. 5). With 16.8 in. I.D. column, the secondary area is not significant for normal to fine packings, but it starts to play an important role for course packings (Fig. 2). Since the wall area is dominant in the secondary area for the whole spectrum of packing a_P, top and bottom area have been omitted in the future correction of secondary effect for simplification. The column diameter will greatly affect the $a_{secondary}$ as expected (Fig. 3). Therefore, this correction will enable wider application of the area model to column with different sizes. Different regimes of packing area for MG 64Y have been shown in Fig. 4 for illustration.

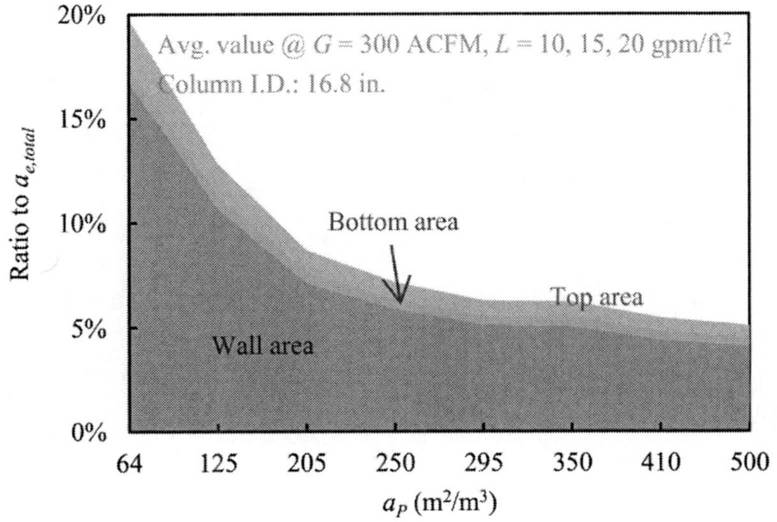

Fig. 2: Effect of a_P on $a_{secondary}$

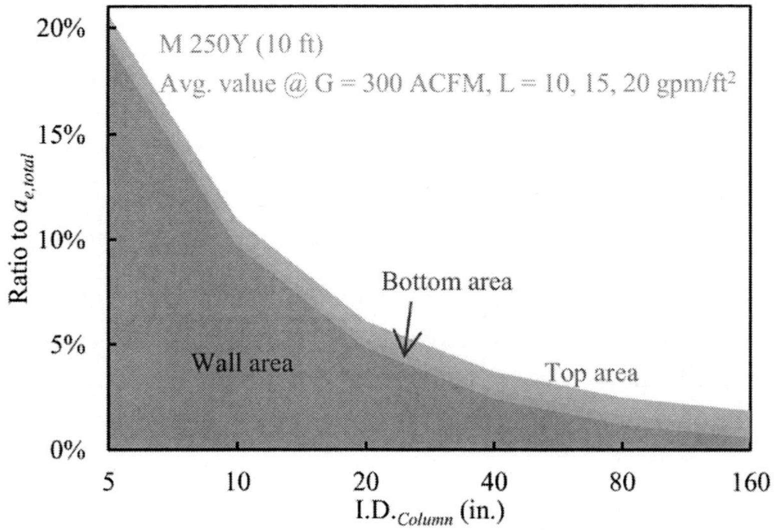

Fig. 3: Effect of column size on $a_{secondary}$

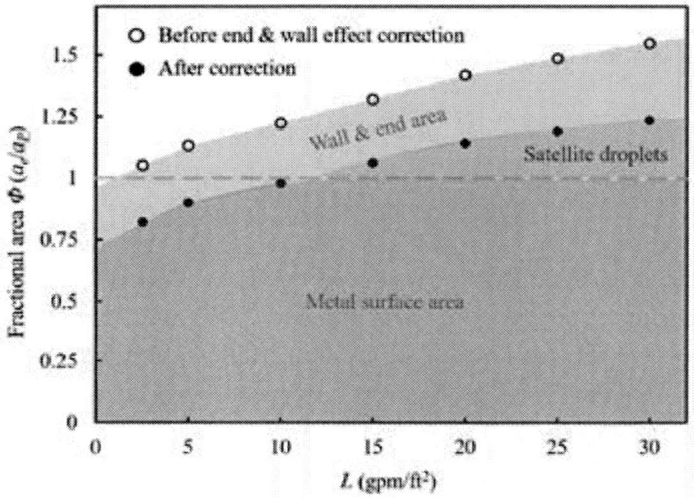

Fig. 4: Packing area regimes of MG 64Y

Figure 5 compares the measured and calculated area with water. An empirical correction factor has been included in the model to account for deviation of random/hybrid packings and plastic packings (Eq. 6 and Tab. 1). For 27 packings with drastically different geometry, the model shows an average absolute deviation (AAD) of only 6.3 percent (Fig. 6).

$$\eta = \eta_{type} \cdot \eta_{material} \tag{6}$$

Tab. 1: Packing correction factors for area model

η_{type}	Structured	Random/Hybrid
	1.0	$-2\times10^{-5}a_P^2+7\times10^{-3}a_P+0.64$
$\eta_{material}$	Stainless Steel	Plastic
	1.0	0.7

Per the model, surface tension affects $a_{e,packing}$ but not viscosity. This is confirmed by Fig. 7 & 8 where no bias of error was observed as a function of the two physical properties. For viscosity, both solutions of polyethylene glycol (PEG) and glycerol have shown no difference in the $a_{e,packing}$ compared to water.

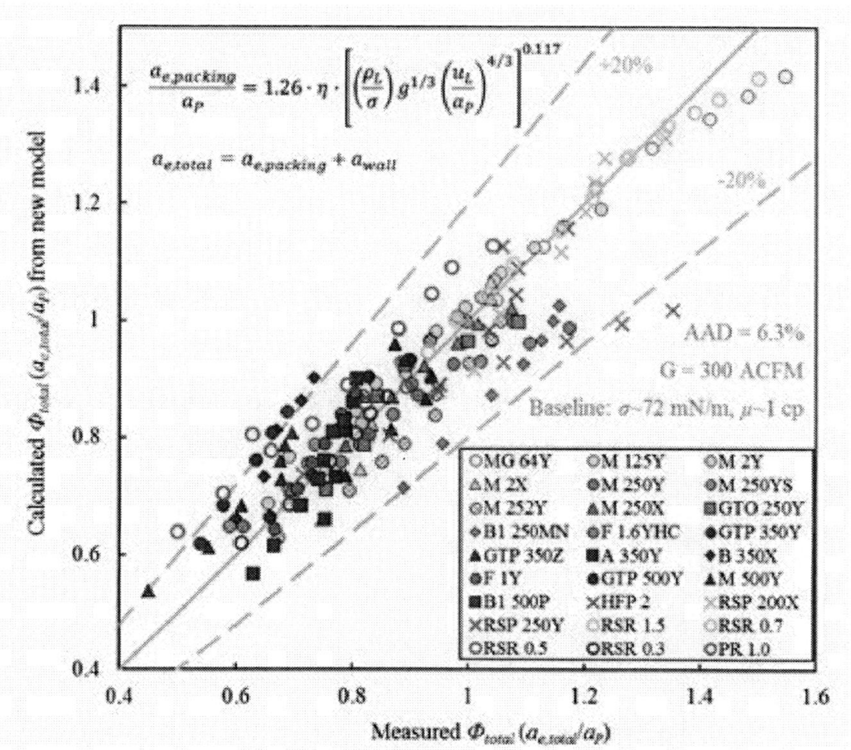

$$\frac{a_{e,packing}}{a_p} = 1.26 \cdot \eta \cdot \left[\left(\frac{\rho_L}{\sigma} \right) g^{1/3} \left(\frac{u_L}{a_p} \right)^{4/3} \right]^{0.117}$$

$$a_{e,total} = a_{e,packing} + a_{wall}$$

Fig. 5: Comparison of measured and calculated area from Eq. 1 for water

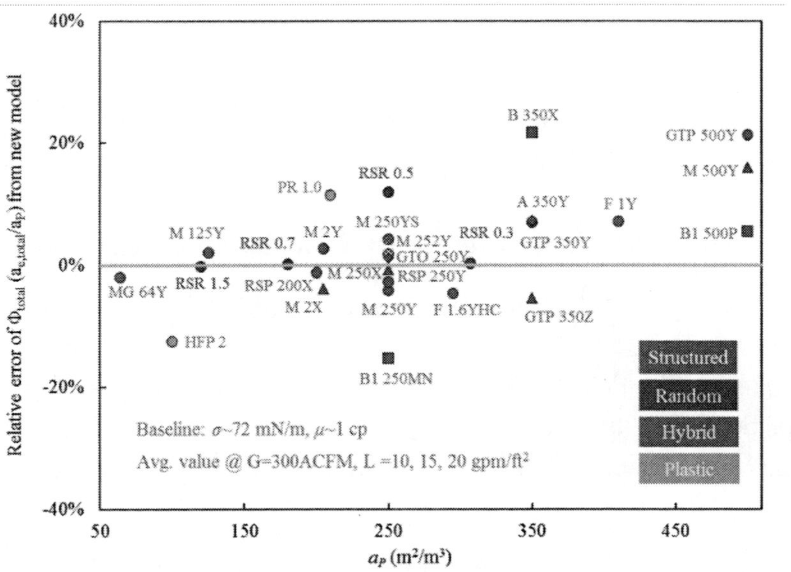

Fig. 6: Relative error of Eq. 1 for different packings

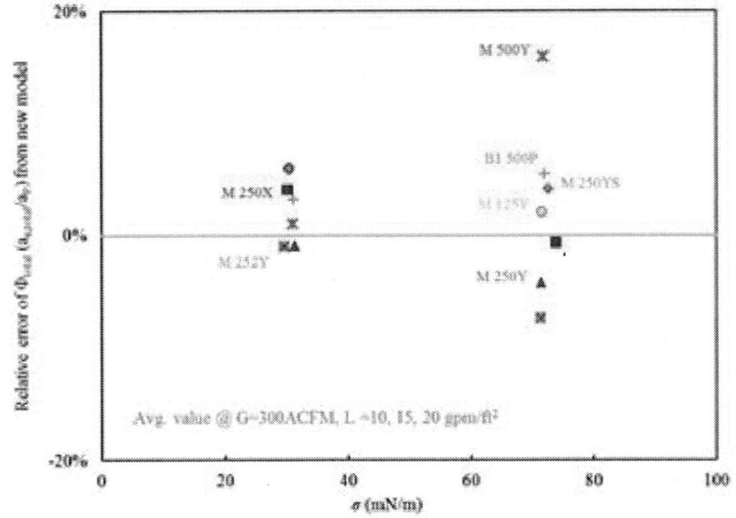

Fig. 7: Relative error of Eq. 1 for different surface tension

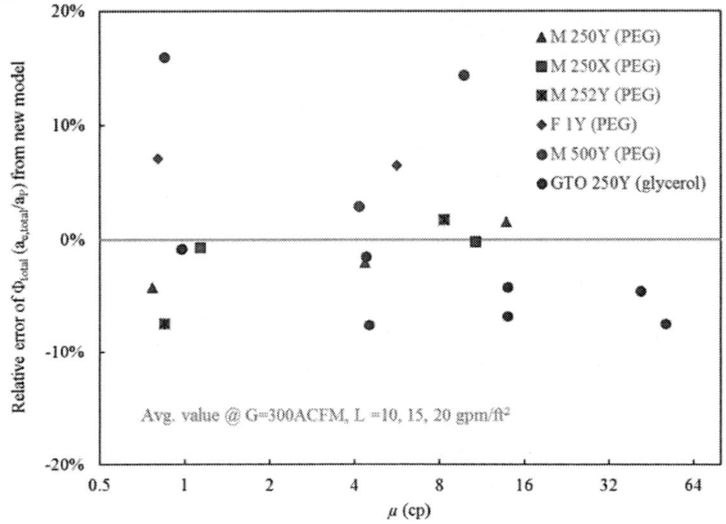

Fig. 8: Relative error of Eq. 1 for different viscosity

Model of Liquid Film Mass Transfer Coefficient (k_L)

The experimental protocol for k_L measurement of SRP has been changed over time. A total of 10 10 ft of packing was initially used. The height was then reduced to 6 ft due to the low outlet toluene concentration that is close to the detection limit of the gas chromatograph (GC). Theoretically, k_L should not be affected by packing height when there is no issue with liquid distribution and GC detection. However, greater k_L has been observed with the 6 ft bed than the 10 ft bed, regardless of packing type and experiment operators (Fig. 9).

The difference of k_L with different bed height may be caused by both the poorer liquid distribution and by issue with the GC detection limit at greater packing bed height. In the development of k_L model, k_L measured with 10 ft bed will be multiplied by 1.32 to be able to compare with those measured with 6 ft bed.

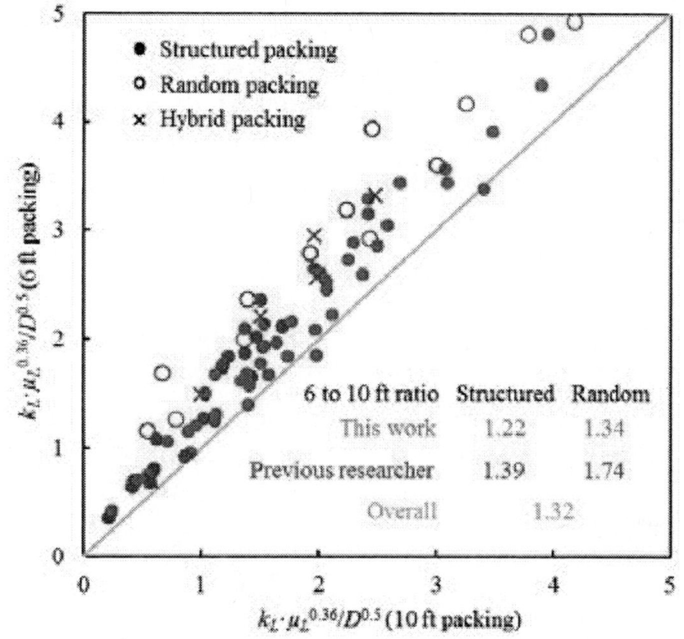

Figure 9: Comparison of k_L measured with 10 ft and 6 ft packing

The average k_L of structured, random, and hybrid packings at 300 ACFM gas rate and 10, 15, and 20 gpm/ft^2 liquid load have been plotted against a_P (Fig. 10). For structured packings, there is a counter-intuitive trend that the k_L decreases with increasing a_P. This might be caused by the longer flow path (and thus greater turbulence) of the liquid on the surface of course packings. The corrugation angle does not have a clear effect on k_L.

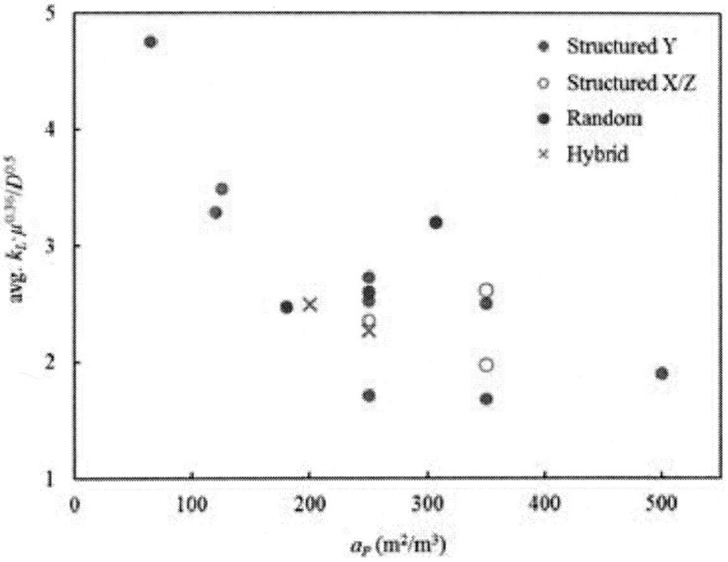

Fig. 10: Average k_L as a function of a_P

Figure 11 compares the measured k_L and calculated k_L from Eq. 2 for 21 packing. Glycerol and water tests have been performed on nine of the packings, and water tests only on the rest of 12 packings. The new model shows an AAD of 20% for all 21 packings. Unlike the area model, the k_L model does not require a correction for random or hybrid packings.

Figure 12 shows the total dependence on μ_L of k_L. The slope (-0.72) is the same for structured, random, and hybrid packings, which means the impeding effect of μ_L on k_L is the same for all three types of packings. Of the total -0.72 dependence, -0.35 is from the indirect influence of μ_L through diffusivity, and -0.37 is from the direct influence of μ_L on k_L through liquid turbulence. The indirect part is believed to be system dependent, and requires knowledge on the D-μ relationship. The direct part is universally applicable to all Newtonian liquids.

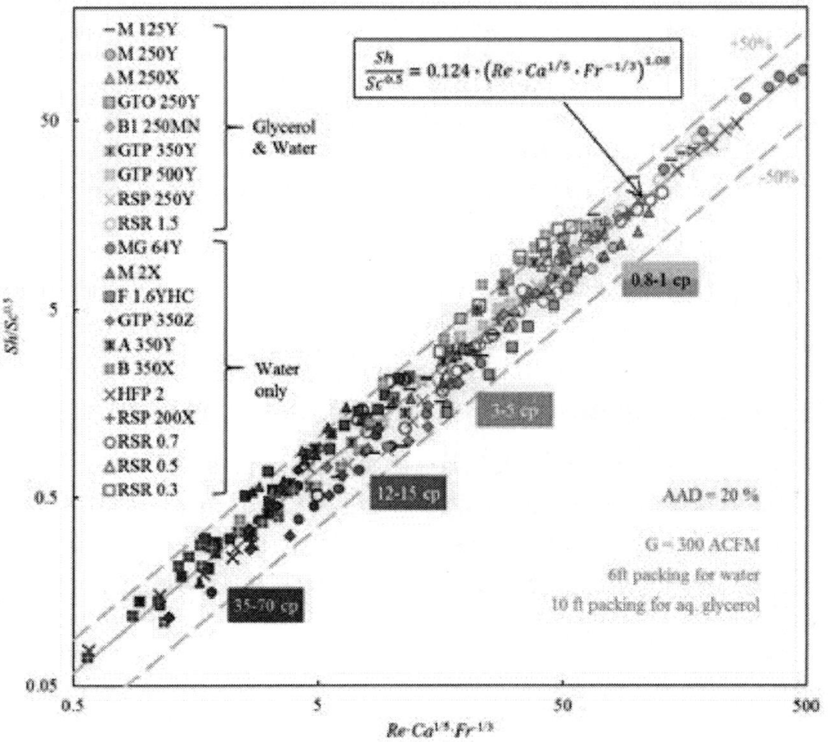

Fig. 11: Comparison of measured and calculated k_L from Eq. 2

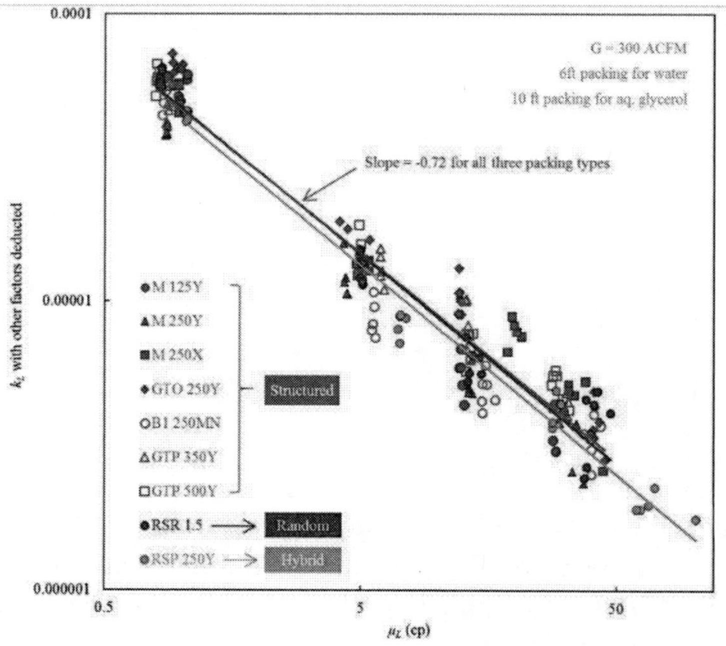

Fig. 12: Total dependence on μ_L of k_L

Figures 13 & 14 shows the relative error of Eq. 2 as function of a_P, and μ_L, respectively. No bias of error was observed as a function of a_P for water in Fig. 13. The model underpredicts k_L for water slightly. In Fig. 14, no bias of error was observed as a function of μ_L.

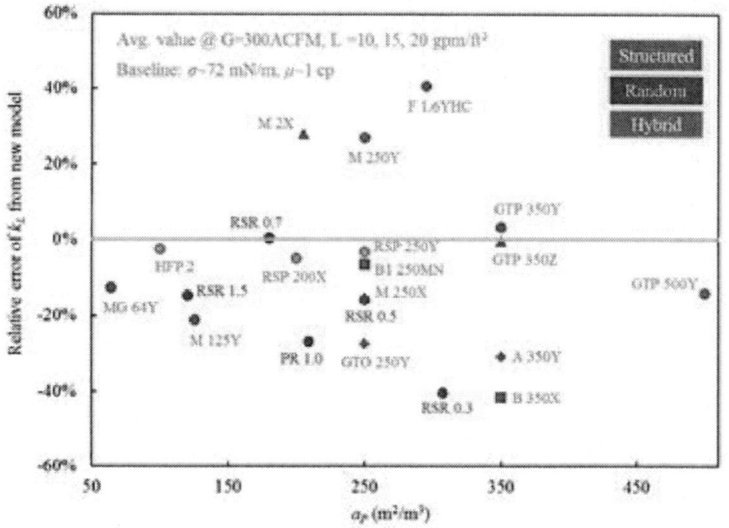

Fig. 13: Relative error of Eq. 2 for different a_P

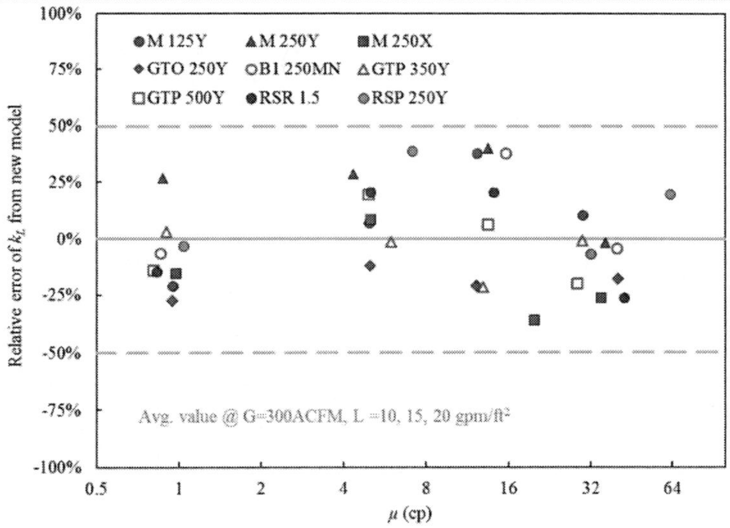

Fig. 14: Relative error of Eq. 2 for different μ_L

Conclusions

Models of a_e and k_L (Eq.s 1 & 2) have been developed based on the SRP air-water column database with 27 packings of various type, a_P, and materials. Liquid viscosity has been varied from 0.8 to 70 cP by the addition of glycerol. The μ_L does not affect $a_{e,packing}$, but it affects k_L with a total dependence of -0.72, of which -0.35 is from the indirect influence through diffusivity, and -0.37 is from the direct influence of μ_L on k_L through liquid turbulence.

The wall effect is dominant in the secondary effect on a_e, and it has been included in the a_e model. The k_L decreases with increasing a_P for structured packings, and shorter packing bed consistently shows greater k_L than taller bed with the same packing internal. With a relatively large equipment size and packing database, this work has the greatly expanded the data of mass transfer in viscous liquids available in the open literature.

Nomenclature

a_e – effective area of mass transfer per volume of column, m^2/m^3
a_P – specific area of packing per volume of column, m^2/m^3
C – concentration, ppmv or ppmw
D – diffusivity, m^2/s
d – characteristic length in dimensionless groups (liquid film thickness in a_e model, and $1/a_P$ in k_L model), m
g – gravitational acceleration, m/s^2
k_g' – liquid film mass transfer coefficient with chemical reactions expressed in gas unit, $mol/m^2 \cdot Pa \cdot s$
k_L – liquid film mass transfer coefficient, m/s
K_{OG} – overall mass transfer coefficient expressed in gas unit, $mol/m^2 \cdot Pa \cdot s$
u – characteristic velocity in dimensionless groups (liquid film velocity in a_e model, and liquid superficial velocity in k_L model), m/s
Z – packing height

Greek letters

σ – surface tension, N/m
μ – viscosity, cP
ρ – density, kg/m^3
η – correction factor for area model

Dimensionless groups

Ca – Capillary number, $\mu \cdot u/\sigma$
Fr – Froude number, $u^2/g \cdot d$
Re – Reynolds number, $d \cdot u \cdot \rho/ \mu$
Sc – Schmidt number, $\mu/\rho \cdot D$
Sh – Sherwood number, $k_L \cdot d/D$
We – Webber number, $\rho \cdot u^2 \cdot d/\sigma$

Packing names

B1 – MONTZ-Pak Type B1TM
F – Flexipac®
GTO – GT-OPTIMPAKTM
GTP – GT-PAKTM
HFP – Hiflow® Plus
M – MellaPakTM
MG – MellaGridTM
PR – Pall Ring

RSP – Raschig Super-Pak®
RSR – Raschig Super-Ring®

References

1. Song, D., A.F. Seibert, and G.T. Rochelle, *Effect of Liquid Viscosity on the Liquid Phase Mass Transfer Coefficient of Packing.* Energy Procedia, 2014. **63**(0): p. 1268-1286.
2. Tsai, R.E., *Mass Transfer Area of Structured Packing*, in *Department of Chemical Engineering.* 2010, The University of Texas at Austin: USA.
3. Wang, C., *Mass Transfer Coefficients and Effective Area of Packing*, in *Department of Chemical Engineering.* 2015, The University of Texas at Austin: USA.
4. Pohorecki, R. and W. Moniuk, *Kinetics of reaction between carbon dioxide and hydroxyl ions in aqueous electrolyte solutions.* chemical Engineering Science, 1988. **43**(7): p. 1677–1684.
5. Song, D. and G.T. Rochelle, *Reaction kinetics of carbon dioxide and hydroxide in aqueous glycerol.* Chemical Engineering Science, 2017. **161**: p. 151-158.

480985 Novel Solvent Exchange Distillation Process- a Comparison Study with Batch Systems

Abstract:

A novel solvent exchange distillation process for both zeotropic and azeotropic equilibrium, which has been developed for replacement of one solvent with a second solvent while keeping the concentration of dissolved nonvolatile solute constant, is described. Detailed simulations for methanol or ethanol replaced by water show that solvent to feed ratios near 1.0 are often sufficient. Comparison of the results with detailed batch distillation simulations show that the novel solvent exchange method gives the same purity with lower solvent and energy consumption. When the less volatile solvent has to be replaced with a more volatile solvent, batch distillation, which requires very large amounts of added solvent and of energy, would normally not be used. Thus, the new solvent exchange distillation process represents a totally new approach to this exchange problem.

AUTHOR INDEX

AIChE
120 Wall Street, FL 23
New York, NY 10005-4020

ISBN 978-1-5108-4137-6

CURRAN ASSOCIATES INC.
proceedings
.com